Climate

Miracle

Foreword by Jerry Molen

There is no climate crisis

Nature controls climate

Ed Berry, PhD

Climate Miracle

There is no climate crisis. Nature controls climate.

Edwin X Berry, PhD, Physics
Certified Consulting Meteorologist

Ed Berry, LLC
439 Grand Dr #147
Bigfork, Montana 59911
ed@edberry.com

Dr. Berry speaks, teaches, advises, and consults on climate change.

Go to:

https://edberry.com/climate-miracle/

to:

- See the References for this book
- Join the conversation about this book
- Join the *Climate Miracle* Membership

Copyright

Climate Miracle:

There is no climate crisis. Nature controls climate.

Dedication

This book is dedicated to all true climate scientists and people everywhere who understand that nature controls the climate,

That the United Nations climate theory "is the greatest and most successful pseudoscientific fraud" in history,

And that truth will prevail if we continue to fight for the truth.

Table of Contents

Foreword

The newly acquired knowledge I have about a new book from Dr. Ed Berry is almost more than I can bear.

As an octogenarian I am nearly overwhelmed with the discovery that truth is once again in season.

And now the world will gain, along with me, a reinforcement of values lost, of twisted information long disseminated by men and women of letters and scoundrels bent on changing our beautiful world to a place of haunted and broken values meant to deceive and corrupt the populace for reasons only they can conceive and believe.

Dr. Berry has brilliantly presented a rare and much needed look at climate change through a prism of honest evaluation, scientific precepts and factual data that finally, yes, finally portrays the truth about what is happening, or not, to our planet in the stars.

Ready or not, you will like "Climate Miracle" because it allows one simple fact that Dr. Berry has exposed….'truth is now in season'.

And now the world knows…….

Gerald R. Molen

Academy Award winning Producer of Schindler's List, Jurassic Park, Hook, Rain Man, Minority Report, and many others.

2

Introduction

They told you they would save the Earth if you followed their instructions. You must believe what they tell you, accept higher energy costs, pay much more in taxes, forget your personal freedom, and vote for their globalist world government. Then they will absolve you from your environmental sins and save the planet for you.

The cornerstone of their message is only government can solve your climate problem. That should be a warning to you that they are lying.

Climate Miracle is not a treatise on climate science. I won't waste your time with unnecessary science.

Climate Miracle gives you the information you need to checkmate the alarmist claim that human CO_2 threatens the planet.

Climate Miracle shows why the alarmists' claim is a fraud.

The *Climate Miracle* is nature controls the climate. Human emissions do not overpower the natural forces that drive an ever-changing climate.

> **Alarmists die a thousand deaths.**
>
> **The wise die only once.**

Chapter 1 – A Critical Climate Debate

"We've got to ride the global-warming issue. Even if the theory of global warming is wrong, we will be doing the right thing, in terms of economic policy and environmental policy." - Timothy Wirth, Clinton Administration Under Secretary of State

"Consensus is the business of politics. If it's consensus, it isn't science. If it's science, it isn't consensus. Period." - Michael Crichton, scientist, physician, author

Let's drop in on a climate debate that never happened.

The fictitious debate is between Globalist Gore and Wisdom Will. It illustrates a climate debate without going deep into the science.

The key reference in this book is the United Nations (UN) Intergovernmental Panel on Climate Change (IPCC).

The Moderator asks Globalist Gore about climate change.

1.1 There are two kinds of climate scientists.

Globalist Gore

More than 16,000 scientists from 184 countries published a letter in 2017, warning that "human beings and the natural world are on a collision course."

More than 11,000 researchers from around the world have issued a grim warning of the "untold suffering" that will be caused by climate change if humanity doesn't change its ways. These scientists say they have the "moral obligation to tell it like it is."

They say posterity will remember them badly if they dismiss climate change as a serious threat to our civilization.

Wisdom Will

There are two kinds of climate scientists: "cause" scientists and "effects" scientists.

All the scientists you reference are "effects" scientists. They focus on the effects of climate change. They tell you climate change causes bad stuff to happen. Then they scare you into believing humans caused the bad stuff.

Their science error is they assume, incorrectly, that human CO_2 causes all the CO_2 increase. They have no idea their core assumption is wrong.

"Cause" scientists focus on the cause of climate change. They are the physicists. And they have proved human CO_2 has little effect on atmospheric CO_2. (Section 2.4)

The idea that 97 percent of scientists support Globalist Gore is false, and even if it were true it would be irrelevant to science.

According to the scientific method, one proof that a theory is wrong outvotes all scientists who claim the theory is true. (Section 5.2)

There is no climate crisis. Nature controls climate.

1.2 Ice core data underrepresent actual CO_2.

Globalist Gore

Ice core data prove natural CO_2 stayed at 280 ppm. Therefore, human CO_2 caused the increase in atmospheric CO_2 above 280 ppm.

Wisdom Will

First, according to the scientific method, it is impossible to prove a theory is correct, but it is possible to prove a theory is wrong.

Second, ice core data underrepresent the true CO_2 level. Leaf stomata data and more than 90,000 direct chemical measurements

6

show the CO_2 level in the past 1200 years was well above your assumed 280 ppm. (Sections 3.4 and 3.5)

1.3 Effects do not prove their cause.

Globalist Gore

On October 4, 2020, "60 Minutes" covered the California fires and concluded human CO_2 was responsible for the dry weather and dead trees that helped the fires to spread so fast.

These wildfires prove our CO_2 is causing dangerous climate change. Floods, droughts, hurricanes and rising oceans are worse than ever.

We must make drastic cuts in our CO_2 emissions. We must add taxes on carbon fuels that will make them more expensive than alternatives. If we don't stop climate change, we will all perish.

The planet is warming. We caused it. We must fix it.

Wisdom Will

The "60 Minutes" show was junk science. It showed examples of weather effects.

Their expert witnesses were "effects" scientists, not "cause" scientists.

One witness claimed California's current drought is a megadrought that occurs only once in a thousand years. But data show California had similar droughts in 1840 and 1930.

More importantly, these droughts are cyclical. They show up in tree-ring data. A 1990 study found cyclical patterns in tree-ring data that predicted a severe drought in 2020 to 2030.

California's drought is a result of natural climate cycles and has nothing to do with CO_2. But California could have minimized its fires by clearing out dead wood and brush. (Section 6.7)

Nature controls the climate. We are not responsible. We can't fix it.

7

Moderator

Sorry everyone. We must take a ten-minute break because about 30 people in the audience have fainted and four appear to have heart attacks. We need time to get them medical attention.

... OK, we are now ready to continue this debate. It's your turn, Globalist Gore.

1.4 Human CO_2 does not stick in the atmosphere.

Globalist Gore

Our CO_2 sticks in the atmosphere for thousands of years. Our CO_2 sticks in the atmosphere like our garbage sticks in a garbage dump.

We must stop our CO_2 emission no matter how much it costs us.

Wisdom Will

Human CO_2 does not stick in the atmosphere. It flows out of the atmosphere as natural CO_2 flows out of the atmosphere. (Section 4.1)

Human and natural CO_2 will behave the same because their CO_2 molecules are the same. If natural CO_2 sticks in the atmosphere for thousands of years, then the CO_2 level would be over a million ppm. (Section 2.8)

Since that has not happened, no CO_2 sticks in the atmosphere for thousands of years.

1.5 Sum of human CO_2 proves nature caused the increase.

Globalist Gore

The sum of human CO_2 emitted since 1750 is **greater than** the increase in CO_2 above 280 ppm. This proves human CO_2 caused the increase.

Wisdom Will

8

IPCC's own data show your claim is wrong. The "sum of human CO_2" is **less than** the "increase in CO_2 above 280 ppm" before 1950. This proves natural CO_2 caused the CO_2 increase. (Section 3.3)

1.6 Statistics prove nature caused the increase.

Globalist Gore

Since 1750, the CO_2 level has increased as human CO_2 emissions increased. This proves human CO_2 caused the increase.

Wisdom Will

Globalist Gore, you should read a book on how to be fooled by statistics. The CO_2 data are time series data.

The time-series trends of hemlines of New York models once correlated with the level of Lake Titicaca in the Andes. Which was the cause, and which was the effect?

There are hundreds of examples of time-series correlations that do not have any cause-effect relationship.

Statisticians detrend time-series data before doing a correlation. They have proved the correlation of annual human CO_2 emissions with the annual changes in CO_2 is zero. Zero correlation means human CO_2 is not the cause of the increase in CO_2. (Section 3.6)

Moderator

OK, OK. We must take another break. We have a few more heart attacks in our audience and a breaking riot.

... OK, we are now ready to continue this debate. It's your turn, Globalist Gore.

1.7 People believe human CO_2 caused the increase.

Globalist Gore

Most people know climate change is real. They support aggressive climate legislation to address the crisis.

Wisdom Will

Most people "know climate change is real" because our media, government, schools, colleges, and universities have indoctrinated our people for two generations. They scared them to make them believe climate fiction. Some children were so scared they committed suicide.

Their scary predictions never come true. But the scared people never acknowledge that fact.

They make children join groupthink programs. They teach them to reject facts that contradict their groupthink belief. They taught them to ignore or attack those who opposed their belief.

They claim to be on the side of science, but they promote the false idea that the children they exploit for political purposes are climate experts.

That is not teaching. That is child abuse and brainwashing.

1.8 IPCC's carbon cycle shows nature caused the increase.

Globalist Gore

When you calculate how human carbon flows through the carbon cycle, you will find that human carbon that flows out of the atmosphere flows back into the atmosphere. This backflow causes human CO_2 to cause all the rise in atmospheric CO_2.

Wisdom Will

That is not true, Globalist Gore.

Dr. Ed Berry, an atmospheric physicist, out in Bigfork, Montana, calculated what IPCC scientists did not, or could not do. [This is called a shameless plug – exb.]

He had no government funding like your fat-cat scientists because our government won't fund research that contradicts the IPCC. Berry did the calculations using Microsoft Excel on his desktop computer. He says a good high school student can do the same calculations.

Then, Dr. William Happer, of Princeton University, and W.A. van Wijngaarden, of York University, Canada, proved Dr. Berry's numerical calculations are correct. Many other scientists have reviewed and approved Berry's work.

Berry showed that a few years ago, when the CO_2 level was 413 ppm, natural CO_2 was responsible for 380 ppm and the human effect was only 33 ppm, based on IPCC data.

He also showed if human CO_2 emissions stopped in 2020, the human effect would fall to 10 ppm by 2100. (Section 4.7)

Nature controls CO_2 and there is no climate crisis.

1.9 We don't control the climate.

Globalist Gore

Even if we are wrong about climate science, we must do everything we can to reduce our CO_2 emissions.

Wisdom Will

Economist Bjorn Lomborg, in his book *False Alarm*, shows that no amount of money spent on trying to control global temperature would have any measurable effect. He says adjusting to climate change is more economical and moral than trying to prevent climate change. And he even believes the IPCC climate myth.

Lomborg shows that for less than $100 billion per year – a fraction of the amount the alarmists want to spend on climate change – we can lift the world's 650 million extremely poor people out of their extreme poverty.

Good science shows that nature controls CO_2. We are not responsible for the increase. There is no climate crisis.

Chapter 2 – Climate Change Basics

"No matter if the science is all phony, there are collateral environmental benefits... Climate change provides the greatest chance to bring about justice and equality in the world." - Christine Stewart, former Canadian Environment Minister

"Members of our species simply repeat what they are told – and become upset if they are exposed to any different view." – Michael Crichton, scientist, physician, author

2.1 Define climate change

Climate change means that climate changes. There are two kinds of climate change:

- nature-caused climate change
- human-caused climate change

Climate alarmists use "climate change" to mean "human-caused climate change" because they want you to think you caused the climate change.

2.2 Define the units to measure carbon

You are familiar with speed limits in miles per hour (mph) or kilometers per hour (km/h). To measure anything, we must define the units to be measured.

We measure the level of carbon dioxide or CO_2 in the atmosphere in parts per million by volume (ppmv). However, it is customary to omit the "v" and just write ppm.

We measure the level of carbon without the two oxygens in petagrams (PgC). We can convert CO_2 in ppm into carbon in PgC by multiplying by the ppm by 2.12. For example, if the CO_2 level is 415 ppm, we multiply by 2.12 to get 880 PgC.

2.3 IPCC's three theories

The IPCC makes three fundamental claims or hypotheses. For simplicity, we use the word "theory" to include what scientists would call a "hypothesis."

Figure 2.1 shows IPCC's three connected theories:

1. IPCC's core theory: Natural CO_2 stays constant at 280 ppm, or human CO_2 causes all the increase in atmospheric CO_2.

2. IPCC's second theory: CO_2 increase causes global warming.

3. IPCC's third theory: Global warming causes bad stuff to happen.

IPCC's complete theory is human CO_2 causes dangerous climate change. To support its complete theory all three theories must be true. If any one of its three theories is not true, then IPCC's complete theory is not true.

We show IPCC's core theory is false, which is sufficient to show IPCC's complete theory is false.

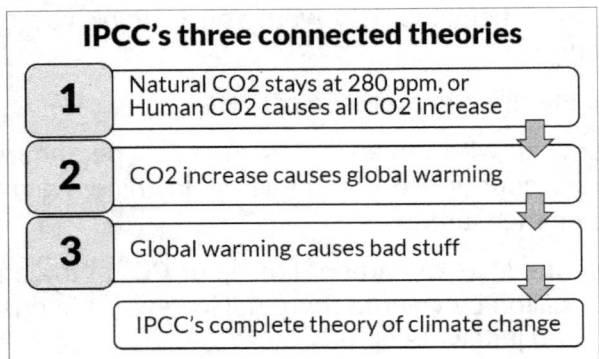

Figure 2.1. The IPCC needs three connected theories to support its complete theory of climate change.

Other books and scientific papers show IPCC's second and third theories are also false.

2.4 There are two kinds of climate scientists

There are two kinds of climate scientists: "cause" scientists and "effects" scientists.

The "cause" scientists focus on IPCC's first two theories. They are the physicists who work to determine the causes of climate change. They also include some chemists, geologists, and smart self-educated people. But the cause-effect relationship is rooted in physics.

The "effects" scientists assume IPCC's first two theories are true. They focus on the effects of climate change under the assumption that human CO_2 causes the climate change.

Who are the "cause" experts? The "cause" scientists.

Who are the "effects" experts? The "effects" scientists.

The climate world mistakes "effects" scientists for "cause" experts. But they are entirely different.

2.5 IPCC's core theory

The IPCC claims,

> "With a very high level of confidence, the increase in CO_2 emissions from fossil fuel burning and those arising from land use change are the dominant cause of the observed increase in atmospheric CO_2 concentration."

More simply, IPCC's core theory is:

The natural CO_2 level has stayed constant at 280 ppm.

IPCC's core theory predicts, or is equivalent to:

Human CO_2 caused all the increase in atmospheric CO_2 above 280 ppm.

The IPCC argues this is so because:

> Ice-core data show CO_2 remained at 280 ppm for a few thousand years before 1750. Therefore, the natural CO_2 level must have remained at 280 ppm after 1750.

If IPCC's core theory is false, as we prove it is, then all IPCC's climate claims have no scientific basis. The IPCC has no alternative theory to take the place of its core theory. The only alternative theory is the accurate physics model described in Chapter 4.

Climate alarmists, without knowing it, base all their climate claims on their incorrect assumption that IPCC's core theory is valid.

This book focuses on IPCC's core theory because it is easy to prove IPCC's core theory is wrong and this proves all IPCC's climate claims are invalid.

2.6 IPCC's second theory is also wrong.

A prediction connects a cause to an effect.

A long time ago, someone observed that a rooster always crowed before the sun rose. If that someone was a budding scientist, that someone may have proposed the theory that the rooster's crowing made the sun rise from its sleep.

Don't laugh. That theory makes more sense than IPCC's second theory.

Every time they tested the rooster theory, the rooster crowed before the sun rose.

But modern data show temperature increases occur before CO_2 increases and temperature decreases occur before CO_2 decreases. Temperature changes precede CO_2 changes by 80 to 800 years.

Yet, IPCC's second theory says CO_2 changes cause the temperature changes. Go figure.

Furthermore, if higher temperature increases CO_2, as the data show, and if higher CO_2 increased temperature as IPCC claims, then global temperature would run away, and we would not be here.

Yes, scientists have calculated that more CO_2 causes a small amount of warming. But such calculations do not include the entire Earth climate system where feedbacks will occur.

The IPCC admits climate is a complex system. And systems engineering shows that complex systems usually produce results opposite to what simplistic thinking predicts.

There are good science papers that show human CO_2 has no effect on climate at all.

We will not discuss IPCC's second theory more than this in this book because that subject belongs in another book.

2.7 IPCC's third theory is an illusion.

Human civilization developed during our Holocene warm period of the last 12,000 years. The Holocene had several periods that were warmer than now. The preceding ice age of 90,000 years was not a comfortable time to live. Before our Holocene warm period, humans lived like animals.

If you consider the bigger climate picture, you might ask, "Why are we worried about global warming?"

Chapter 3 – IPCC's Core Theory

"But the basic physics of the climate are well understood. Burning fossil fuels emits carbon dioxide. And carbon dioxide is a greenhouse gas that traps heat in the atmosphere." - Christine Todd Whitman, the former EPA chief under President George W. Bush

"Consensus is invoked only in situations where the science is not solid enough. Nobody says the consensus of scientists agrees that $E=mc^2$. Nobody says the consensus is that the sun is 93 million miles away." - Michael Crichton, scientist, physician, author

3.1 The Equivalence Principle for climate

The Equivalence Principle says if data cannot tell the difference between two things then the two things are identical. The Equivalence Principle is a foundation of Einstein's theory of general relativity. It says the inability to distinguish between gravitational and inertial forces means they are the same thing.

The Equivalence Principle applies to climate because nature cannot tell the difference between human-produced and nature-produced CO_2 because their molecules are identical.

Therefore, nature treats human and natural CO_2 molecules the same. Any theory or conclusion or model that violates the Equivalence Principle is wrong.

IPCC's theories and claims violate the Equivalence Principle.

3.2 Human CO_2 does not stick in the atmosphere.

The IPCC claims,

> "The removal of human-emitted CO_2 from the atmosphere by natural processes will take a few hundred thousand years (high confidence)."

The Equivalence Principle shows this IPCC claim is impossible. The Equivalence Principle says human CO_2 flows out of the atmosphere as natural CO_2 flows out of the atmosphere.

If natural CO_2 sticks in the atmosphere for thousands of years, then the CO_2 level would be over one million ppm. So, obviously, natural CO_2 does not stick in the atmosphere.

Therefore, neither human nor natural CO_2 sticks in the atmosphere for thousands of years. Both human and natural CO_2 flow out of the atmosphere equally in only a few years.

3.3 Sum of human CO_2

The IPCC argues that its core theory is true because:

> The sum of human carbon emissions since 1750 exceeds atmospheric carbon, so human emissions caused all the increase in CO_2.

The IPCC argues that human CO_2 caused all the CO_2 increase above 280 ppm as follows:

a) In any one year, the sum of human CO_2 over all previous years is greater than the increase in atmospheric CO_2.

b) Therefore, human CO_2 caused all the CO_2 increase.

The "Sum of human CO_2" is the total of all previous human CO_2 emissions.

Of course, the use of the "Sum of human CO_2" is a strawman because it assumes human CO_2 sticks in the atmosphere, which is untrue. However, even if we allow that invalid IPCC assumption for the sake of this argument, the IPCC still loses.

Figure 3.1 shows the data to test IPCC's argument plotted from 1820 to 2020. Not much happened before 1820.

a) The measured "CO_2 level" (solid line) comes from IPCC-approved ice core data and CO_2 data from Mauna Loa.

20

b) The "Sum of human CO_2 emissions" (dashed line) comes from IPCC-approved human CO_2 annual emissions data.

Figure 3.1 shows with red bars that the "Sum of human CO_2" is **less than** "CO_2 level data" before 1950.

Therefore, human CO_2 cannot be responsible for the rise in the CO_2 level before 1950. Therefore, IPCC's first theory is false.

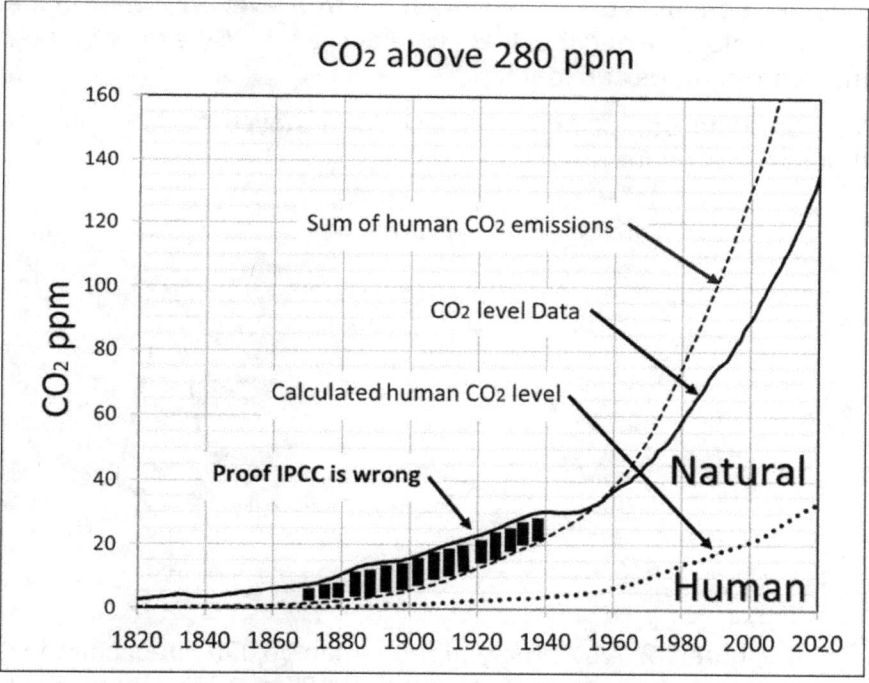

Figure 3.1. The bars show where the CO_2 level is greater than the sum of human CO_2. Therefore, IPCC's core theory is wrong. The "Calculated human CO_2 level" is described in Section 4.7.

This is a checkmate. The climate alarmists' game is over.

Chapter 4 describes how the "Calculated human CO_2 level" is based on IPCC's data.

3.4 Direct CO₂ data prove IPCC's theory is wrong.

Figure 3.2 shows the summary by Beck (2007) of more than 90,000 direct chemical measurements of CO_2. These CO_2 data are well above 280 ppm. They show very high levels in 1820, 1860, and 1940.

The smooth grey line in Figure 3.2 shows the Antarctica ice core data. Ice core data do not represent the true level of CO_2. Both the leaf stomata data and the direct data agree the CO_2 level is greater than shown by ice core data.

Ice core data do not prove the CO_2 level stayed at 280 ppm for thousands of years.

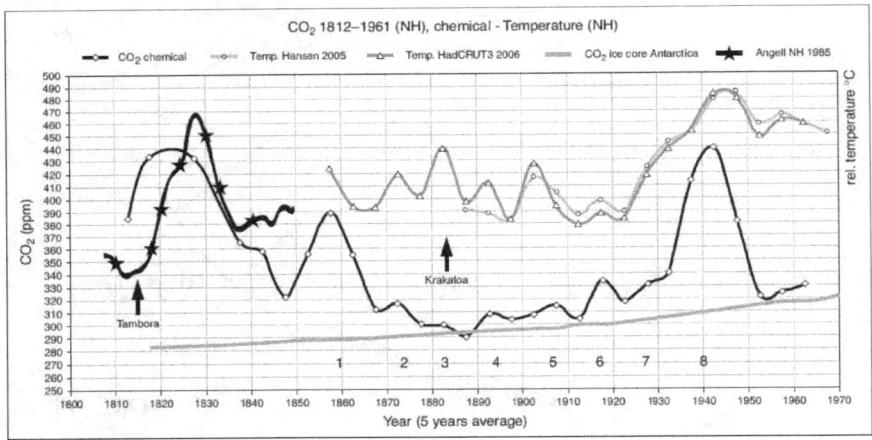

Figure 3.2. A summary of more than 90,000 direct chemical measurements of CO_2 from 1812 to 1961. From Beck (2007) Figure 14.

When a theory contradicts data, the theory is wrong. IPCC's core climate theory is wrong. All IPCC's claims about human-caused climate change are therefore invalid.

3.5 Leaf stomata data prove IPCC's theory is wrong.

Leaf stomata are the leaf pores that control the inflow and outflow of carbon dioxide. The preserved stomata of historical tree leaves provide information of the CO_2 level when the trees were alive.

Figure 3.3 shows the reconstruct historical CO_2 levels conifer leaf stomata data. This is from Kouwenberg (2004). The stomata-derived CO_2 levels are much greater than 280 ppm.

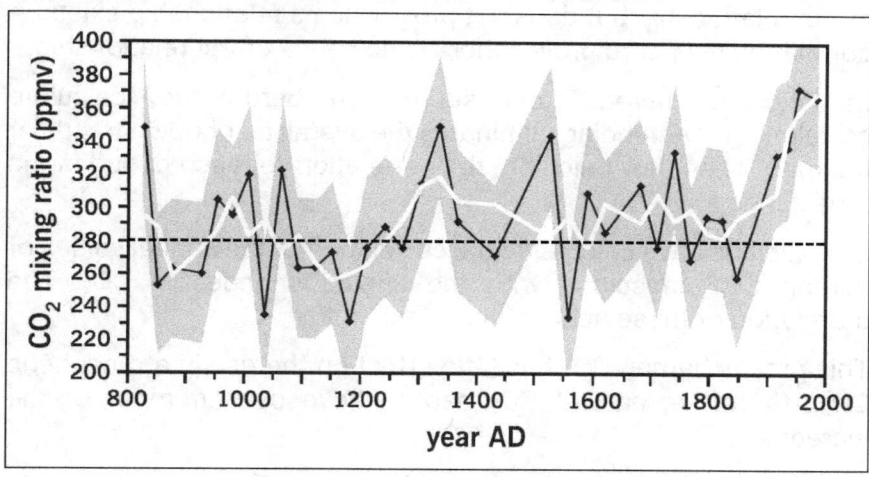

Figure 3.3. Reconstructed CO_2 levels based on stomatal frequency counts. The solid black line is the mean. The white line is a moving average. The grey area shows confidence interval of ± 1 RMSE. The dashed line shows the 280-ppm level. (Kouwenberg, 2004, Figure 4.4).

3.6 Statistics prove IPCC's core theory is wrong.

The IPCC claims its core theory is true because

"... the observed rate of CO_2 increase closely parallels the accumulated emission trends from fossil fuel combustion and from land use changes."

IPCC scientists should read a book on how to be fooled by statistics. The CO_2 data are time series data.

23

There are hundreds of examples of time-series correlations that do not have any cause-effect relationship. So, what is wrong with IPCC's claim?

The purpose of calculating a correlation between two variables is to see if changes in one variable may cause changes in the other variable.

A correlation closer to one than to zero indicates a possible cause-effect relationship but does not prove such a relationship exists. A correlation near zero proves there is no cause-effect relationship.

Statisticians "detrend" time series data before they calculate correlations. Detrending eliminates the overall up or down trends of the data. Then they calculate the correlation for each period in the time series.

Statisticians have calculated the correlation of the annual values of human CO_2 emissions with the annual changes in CO_2. The correlation of these annual data is zero.

This proves human CO_2 has little effect on the rise in atmospheric CO_2. Therefore, natural CO_2 had to have caused most of the increase.

Now, some smart person may say, the correlation is zero because the large variable inflow of natural CO_2 overwhelms the small human inflow. That would be correct but that also admits IPCC's core theory is wrong.

3.7 IPCC's real mission

The IPCC is a political organization, not a scientific organization. Wikipedia provides the following information on the IPCC.

> "The Intergovernmental Panel on Climate Change (IPCC) is an intergovernmental body of the United Nations that is dedicated to providing the world with objective, scientific information relevant to understanding the scientific basis of the risk of human-induced climate change, its natural, political, and economic impacts and risks, and possible response options."

> "The IPCC produces reports that contribute to the work of the United Nations Framework Convention on Climate Change (UNFCCC), the main international treaty on climate change. The objective of the UNFCCC is to
>
> > "stabilize greenhouse gas concentrations in the atmosphere at a level that would prevent dangerous anthropogenic (human-induced) interference with the climate system".

> "The IPCC's Fifth Assessment Report was a critical scientific input into the UNFCCC's Paris Agreement in 2015."

> "The IPCC has adopted and published "Principles Governing IPCC Work", which states that the IPCC will assess:
>
> > • the risk of human-induced climate change,
> >
> > • its potential impacts, and
> >
> > • possible options for prevention."

In summary, the IPCC's mission is to promote, with the help of the UNFCCC, political actions to stop human-caused climate change, even though IPCC's climate theories are invalid.

Chapter 4 – Climate Change Physics

"A global warming treaty must be implemented even if there is no scientific evidence to back the enhanced greenhouse effect." - Richard Benedick, deputy assistant secretary of state, USA

"The climate-change scam, driven entirely by the left, is the world's largest organized-crime fraud." – Lord Christopher Monckton

4.1 The physics model description

CO_2 flows through the atmosphere as water flows through a lake.

Figure 4.1 illustrates a lake. Water from a river flows into the lake and out over a dam. When the outflow equals the inflow the water level will remain constant.

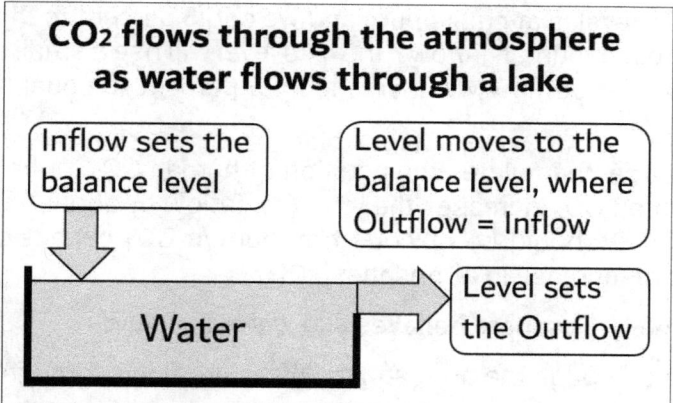

Figure 4.1. A river sets the inflow to a lake. The level of the lake above the dam sets the outflow. Outflow increases as the level increases. The level seeks the level where outflow equals inflow. Then there is no further change in the level as inflow continues.

What do inflow and outflow mean?

Inflow and outflow define how fast water flows into or out of a lake. Flow is expressed in terms of how much stuff flows through a boundary in a certain interval of time.

A river sets the inflow. The height of the level above the dam sets the outflow. The higher the level, the faster the outflow. In other words, the higher the lake level is above the dam, the faster the water will flow over the dam. (The same thing occurs in creeks where water flows over a log or rock.)

If the inflow is constant, the level will adjust until the outflow equals the inflow. No water "accumulates" in the lake. In other words, you don't see a lake level get higher and higher above the level of the dam without more water flowing over the dam. If the inflow goes to zero, the level will fall until it reaches the level of the dam. Then the outflow will be zero.

The lake system always balances. Balance occurs because outflow increases as the level increases.

If the lake level is at equilibrium and we add 5 percent to the inflow – to represent human inflow –the lake level will rise a small amount such that the outflow will increase by 5 percent to equal the new inflow.

This 5 percent simulates the addition of human CO_2 to the natural CO_2. Human CO_2 increases the total CO_2 inflow by about 5 percent. Therefore, it should be obvious that human CO_2 cannot cause a significant increase in atmospheric CO_2.

CO_2 in the atmosphere behaves like water in a lake.

CO_2 flows through the atmosphere like water flows through a lake.

Figure 4.2 illustrates the process. The inflow sets a balance level where the outflow will equal the inflow.

The level sets the outflow. More specifically to the atmosphere and the carbon cycle, the physics model sets the outflow to be directly proportional to the level. If the level doubles, the outflow doubles.

Another way to put it, is to state this as a hypothesis or theory. It goes like this. The physics model theory is:

Outflow is proportional to level.

Notice how different the physics model theory is from IPCC's theories.

The physics model theory states a cause-effect relationship between level and outflow, e.g., if level doubles, outflow doubles. All good theories state a cause-effect relationship.

How CO2 flows through the atmosphere

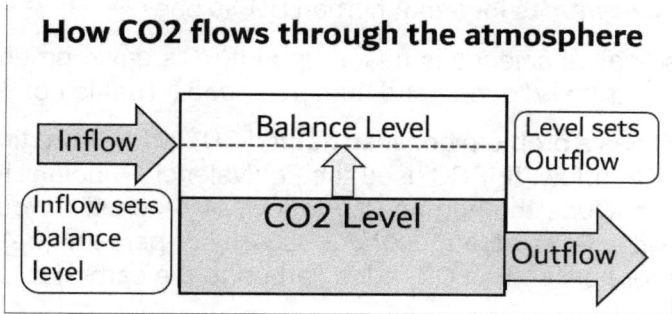

Figure 4.2. How CO_2 flows through the atmosphere.

IPCC's theories state what happens, like the natural CO_2 level stays constant at 280 ppm. The reality is the IPCC needs the physics model to explain how the natural level can stay constant at 280 ppm when the inflow of natural CO_2 is 20 times the inflow of human CO_2.

This is the only theory the physics model needs. With that theory, we can develop all the math we need to represent and calculate a carbon cycle model.

The math in the physics model defines the balance level. The math describes how the level moves to the balance level. If the level is below the balance level, the inflow is greater than the outflow and the level will rise, and vice-versa.

In other words, the system is stable.

The physics model shows why we can calculate the flows of natural carbon and human carbon separately. This simplifies our carbon cycle calculation. We do the calculations separately. We can add them up later.

No model exactly represents nature. The value of a model is how well it represents nature. The physics model uses the simplest possible theory. Incidentally, chemical and pharmaceutical models use the same theory.

4.2 The origin of IPCC's climate alarmism

The IPCC does not use a model like the physics model. Rather the IPCC begins with its idea that human CO_2 is bad.

IPCC's so-called science is based upon IPCC's environmental-cult view that nature is "good," and human is "bad." That is not physics.

The IPCC has a problem because natural CO_2 inflow is 20 times the human CO_2 inflow. If IPCC used the Equivalence Principle, it would have to conclude the human CO_2 inflow, of 5 percent, would only increase the level of atmospheric CO_2 by 5 percent (before we account for how human CO_2 adds carbon to the carbon cycle).

But 5 percent cannot be the problem because natural CO_2 inflow varies by more than 5 percent. So, the addition of human carbon to the carbon cycle does not significantly alter the simple 5 percent argument.

It is difficult for the IPCC to argue that the small human inflow of 5 percent caused all the increase in atmospheric CO_2 above 280 ppm. That "all" is now about 32 percent.

This cult path forces the IPCC to claim that human CO_2 stays in the atmosphere thousands of times longer than natural CO_2 stays in the atmosphere. This violates the Equivalence Principle, but the IPCC ignores that little problem.

Therefore, according to the IPCC, natural CO_2 flows in and out of the atmosphere and its level remains perfectly in balance, as the physics model explains.

But the IPCC needs a magic demon in the atmosphere to capture human CO_2 molecules and make them stay in the atmosphere longer than natural CO_2 molecules stay in the atmosphere.

That IPCC anti-science position is the basis of all climate alarmism.

4.3 IPCC's carbon cycle

Figure 4.3 illustrates IPCC's carbon cycle model. The boxes represent IPCC's four carbon reservoirs: Land, Atmosphere, Surface Ocean, and Deep Ocean.

Each reservoir contains an amount of carbon that we call a "level." The arrows between the reservoirs represent the flows of carbon between the reservoirs.

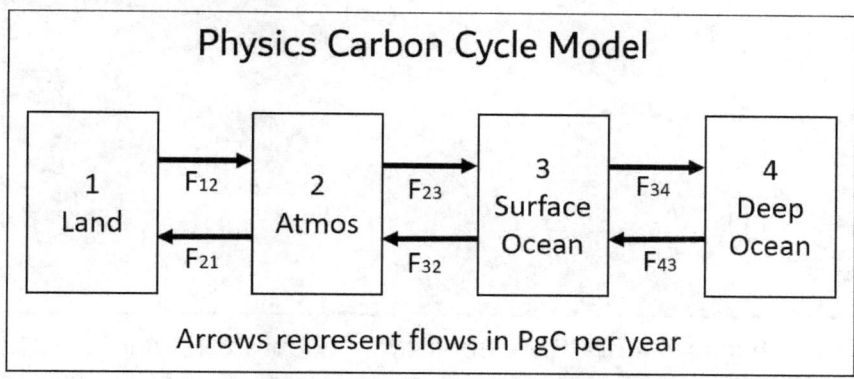

Figure 4.3. Physics model interpretation of IPCC's carbon cycle model.

The physics carbon cycle model uses the 4 levels and the 6 flows in its calculations of how carbon flows from reservoir to reservoir. The levels determine the flows, and the flows change the levels.

Figure 4.4 shows IPCC's illustrated carbon cycle model. The UN IPCC 2013 report, Figure 6.1 shows IPCC's data for the natural and human carbon cycles.

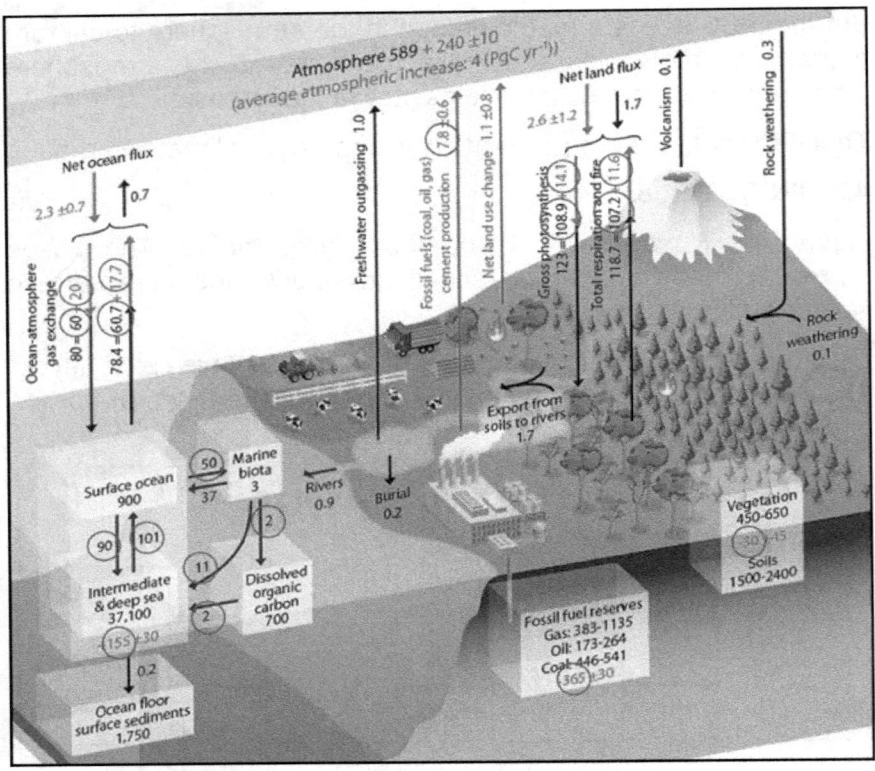

Figure 4.4. IPCC's carbon cycle (Figure 6.1 from its 2013 report). To see the figure in color, please go to the References.

The black numbers represent IPCC's natural carbon cycle. The red numbers represent IPCC's human carbon cycle. The circled numbers are IPCC's flows between the reservoirs.

4.4 IPCC's natural carbon cycle

You are about to learn more about IPCC's carbon cycle than the IPCC knows.

We plot IPCC's reservoir levels as percentages.

Figure 4.5 shows the percentage levels for IPCC's natural carbon cycle. It shows the carbon in the atmosphere is 589 PgC. This is equivalent to 278 ppm of CO_2. This is close to IPCC's core theory that claims natural CO_2 stays at 280 ppm. Therefore, these percentages represent IPCC's natural carbon cycle at equilibrium.

The percentage levels in Figure 4.5 are a "fingerprint" of the natural carbon cycle at equilibrium. According to the Equivalence Principle, the human carbon cycle will have this same fingerprint when it is at equilibrium.

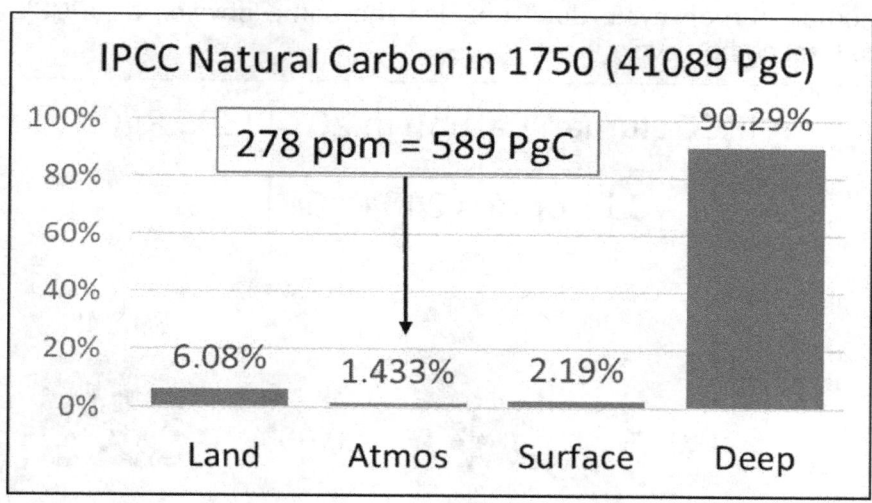

Figure 4.5. Level percentages for IPCC's natural carbon cycle.

Of course, the human carbon cycle will not be at equilibrium so long as human CO_2 flows into the atmosphere. As human carbon flows to the other reservoirs, the human percentage levels move toward the equilibrium fingerprint. And it will do so with the same speed that natural carbon moves to other reservoirs.

Figure 4.5 shows 6.1 percent of natural carbon is in the land, 1.4 percent is in the atmosphere, 2.2 percent is in the surface ocean, and 90 percent is in the deep ocean. These IPCC data are not perfect, but they may be the best data we have on the natural carbon cycle.

4.5 IPCC's human carbon cycle

IPCC's human carbon cycle differs significantly from its natural carbon cycle.

Figure 4.6 shows IPCC's human percentage levels. It shows 61 percent of human carbon in the atmosphere, 39 percent in the deep ocean, and no human carbon in the land or surface ocean.

Simple comparison of Figure 4.6 with Figure 4.5 indicates IPCC's human carbon cycle does not use the same physics as IPCC's natural carbon cycle.

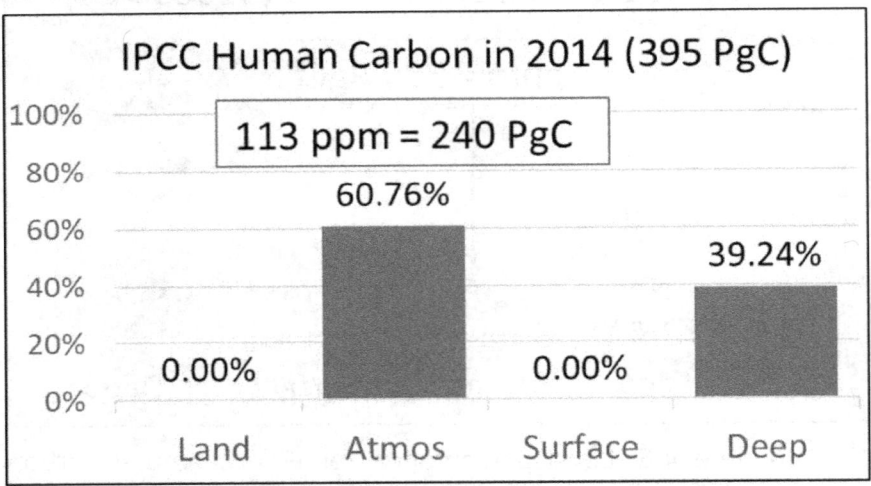

Figure 4.6. Percentage levels for IPCC's human carbon cycle. The 61 percent in the atmosphere results from IPCC's fraud of forcing human CO_2 to cause all the increase in atmospheric CO_2 above 280 ppm.

So, the 61 percent shown for the atmosphere in Figure 4.6 looks suspicious.

Where did the IPCC get its data for its human carbon cycle?

The IPCC did not calculate the distribution of human carbon with a carbon cycle model or there would be carbon in the land and surface ocean reservoirs. Without carbon in the surface ocean, no carbon can flow to the deep ocean.

Rather the IPCC simply inserted into the atmosphere the human carbon it needed to support its core theory and then dumped the remainder in the deep ocean.

This is proof that the IPCC claims human CO_2 caused all the rise in atmospheric CO_2.

The CO_2 level in 2014 was 395 ppm. Subtract 280 ppm to get 114 ppm. This is the CO_2 increase above 280 ppm in 2014. This is also 61 percent of the sum of all human CO_2 emissions through 2014.

Notice IPCC's circular reasoning. It assumes IPCC's core theory is true. Then it inserts the amount of human CO_2 into the atmosphere to satisfy IPCC's core theory. Then it concludes that IPCC's core theory is true. Beautiful circular reasoning.

Granted, the human carbon cycle is not at equilibrium like IPCC's natural carbon cycle. Nevertheless, the human carbon cycle is not too far from equilibrium because human carbon flows from the atmosphere to the other carbon reservoirs.

If human emissions were to stop in 2020, the human carbon cycle percentages would move toward the same percentage levels as IPCC's natural carbon cycle. In fact, we show this in Figure 4.9.

Therefore, the percentage levels for IPCC's human carbon cycle shown in Figure 4.6 should somewhat resemble the percentage levels for IPCC's natural carbon cycle shown in Figure 4.5. But they do not.

We have not done any calculations so far. We have only plotted IPCC's data that the IPCC did not plot. And we found a conflict in the IPCC data that disqualifies IPCC's human carbon cycle.

IPCC's human carbon cycle is the basis of all government policies and laws that claim human CO_2 causes dangerous climate change. This basis is a fraud.

4.6 The IPCC needs a demon for its human carbon cycle

The only way the IPCC can argue that 61 percent of human CO_2 is in the atmosphere while only 1.4 percent of natural CO_2 is in the atmosphere, is to argue that human CO_2 stays in the atmosphere much longer than natural CO_2 stays in the atmosphere.

This IPCC argument requires a demon that can separate human-derived CO_2 from natural-derived CO_2, even though the molecules are identical.

Figure 4.7 illustrates the IPCC demon. The demon separates human CO_2 and natural CO_2 molecules. Then it allows natural CO_2 molecules to flow freely out of the atmosphere.

The demon keeps the human CO_2 molecules in the atmosphere much longer than natural CO_2 stays in the atmosphere. This indeed would be a climate miracle.

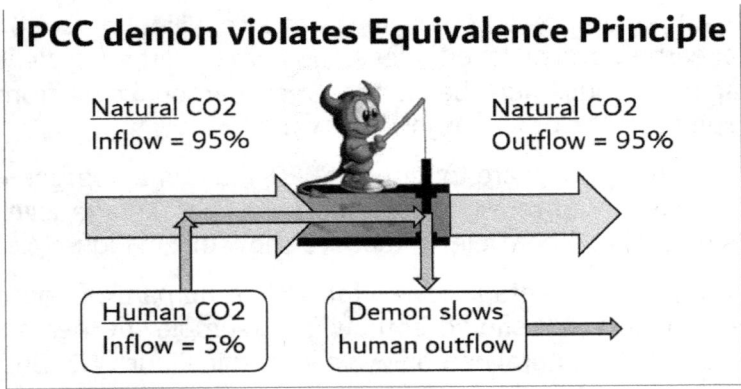

Figure 4.7. The IPCC demon separates human CO_2 from natural CO_2 and slows human CO_2 outflow.

I first presented this demon chart at the "Basic Science of a Changing Climate" Conference in Porto, Portugal, September 7, 2018.

4.7 The physics model human carbon cycle model

This book uses no math. This is the only section that relies on behind-the-scenes math and numerical calculations. The math is shown in my referenced Preprint #3 (see References). It will be published in a scientific journal in June 2021. But you do not need to read or understand the math to understand this book.

Here, we will apply the same physics model to all the flows in IPCC's carbon cycle model shown in Figure 4.2.

This application requires the model to do following:

1. Use IPCC's data for its natural carbon cycle data.

2. Replicate IPCC's natural carbon cycle data.

3. Use IPCC's natural carbon cycle data to calculate the human carbon cycle.

We use IPCC's data for its natural carbon cycle as shown in Figure 4.2 to derive time constants for all six flows. Then we use the same model and time constants to calculate the true human carbon cycle.

I did this calculation with Microsoft Excel. Then, Dr. William Happer, of Princeton University, and W.A. van Wijngaarden, of York University, Canada, showed my numerical calculations are correct.

Figure 4.8 shows the true human carbon cycle percentage levels calculated for 2020.

The human carbon that remains in the atmosphere as of January 2020 is 33 ppm or 15.5 percent, not 61 percent. This means nature added about 100 ppm to atmospheric CO_2 since 1750 according to IPCC's data for its natural carbon cycle.

Figure 4.8 is much closer to the natural carbon cycle fingerprint shown in Figure 4.5, than it is to IPCC's human carbon cycle shown in Figure 4.6.

Figure 3.1 shows the time series, from 1820 to 2020, of this carbon cycle calculation for CO_2 level in the atmosphere.

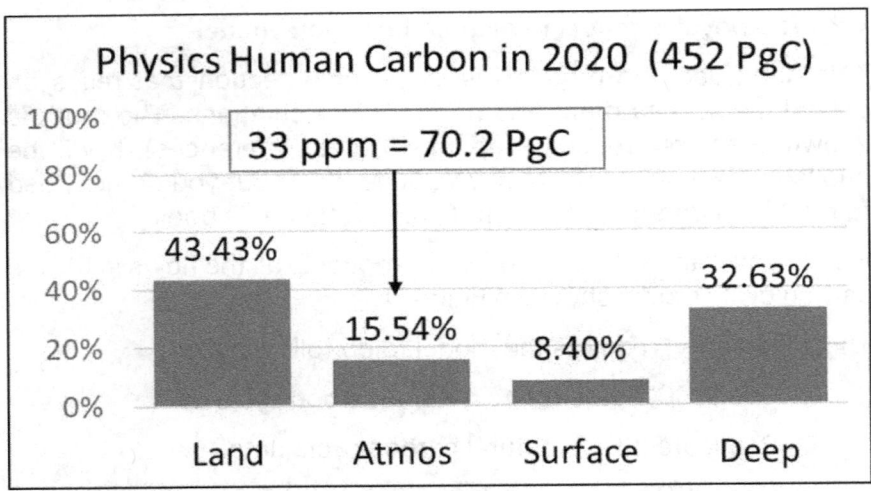

Figure 4.8. The physics model calculation of the human carbon cycle in 2020.

Figure 4.9 shows the percentage levels of human carbon in 2100 assuming all human CO_2 emissions stop on January 1, 2020.

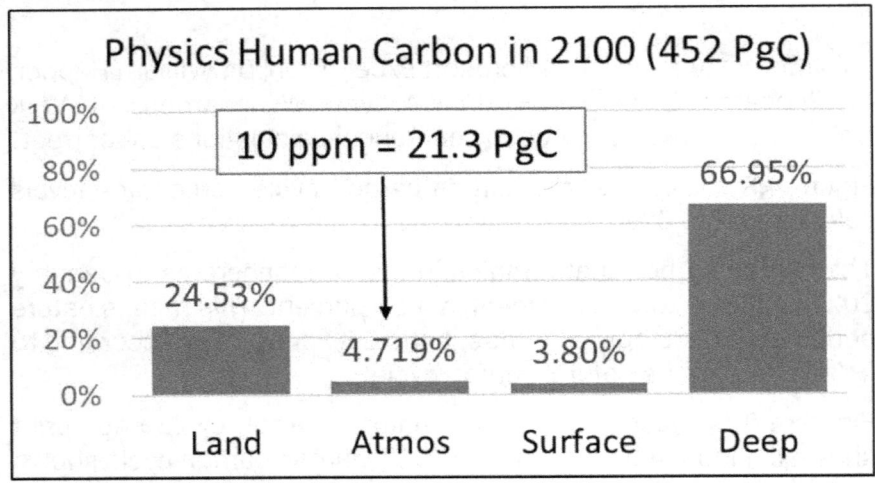

Figure 4.9. The level percentages in 2100 if all human CO_2 emissions stopped in 2020.

The human carbon in the atmosphere would drop from 15.5 percent in 2020 to 4.8 percent in 2100, equivalent to 10 ppm.

The percentages of human carbon in Figure 4.9 are quite close to IPCC's natural carbon percentages shown in Figure 4.5.

Figure 4.10 shows how human CO2 has contributed to atmospheric CO_2 up to 2020. Then the calculation assumes human CO_2 emissions stop in 2020. The calculation shows how fast the human contribution decays from 2020 to 2100. Half of the human component is removed in 20 years.

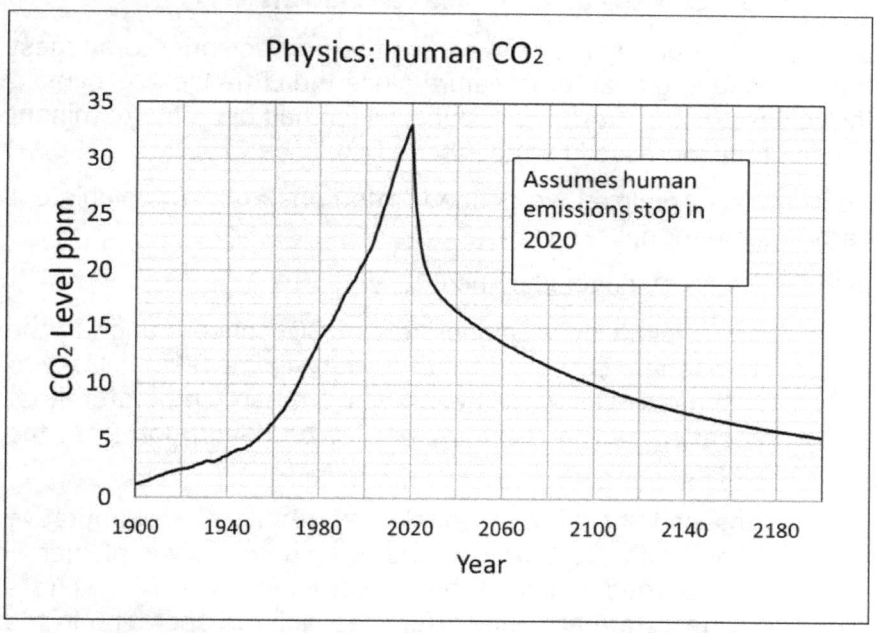

Figure 4.10. Calculated level of human CO_2. Assumes human CO_2 emissions stop in 2020.

These calculations show that human CO_2 emissions are not a long-term threat to the planet.

4.8 Error bounds for these calculations

IPCC says its natural carbon cycle data have 20-percent error bounds. It is easy to insert these error bounds into the physics carbon cycle model.

The corresponding range that human CO_2 may have increased atmospheric CO_2 is from 24 ppm to 48 ppm, according to the physics carbon cycle model. Based upon IPCC's data, human CO_2 levels near these error bounds are very improbable.

4.9 History of these carbon cycle calculations

Dr. Richard Courtney may have been the first scientist to suggest the increase in global temperature since the Little Ice Age caused the release of CO_2 from the oceans, which has been the dominant cause of the increase in atmospheric CO_2 since 1750.

Courtney also realized we needed "rate constants" to calculate a carbon cycle model.

In 2008, Courtney concluded in 2008,

> "... the relatively large increase of CO_2 concentration in the atmosphere in the twentieth century (some 30%) is likely to have been caused by the increased mean temperature that preceded it. The main cause may be desorption from the oceans."

> "Assessment of this conclusion requires a quantitative model of the carbon cycle, but – as previously explained – such a model cannot be constructed because the rate constants are not known for mechanisms operating in the carbon cycle."

In 2019, after reviewing Berry's Preprint #3, Dr. Courtney wrote that Berry's physics carbon cycle model:

> "... quantifies the anthropogenic and natural contributions to changes in atmospheric CO_2 concentration without need for knowledge of rate constants for individual mechanisms.

This is a breakthrough in understanding which [other authors] including myself all failed to make."

In 2008, Courtney correctly realized we need the rate constants to develop a carbon cycle model.

In 2019, Berry realized these necessary rate constants were buried in IPCC's data for its natural carbon cycle. Berry would not have realized that detail had he not previously derived his physics model (Section 4.1) that is based upon the one critical hypothesis that outflow is proportional to level.

It is important to acknowledge that Hermann Harde developed a similar physics model prior to and in parallel with Berry's work. And Harde's work followed the work of Murry Salby.

In 2020, a preliminary paper by Kenneth Skrable, George Chabot, and Clayton French uses carbon 14 data to conclude the dominant cause of the increase in atmospheric CO_2 since 1750 is natural carbon from the ocean and that human CO_2 is a minor cause.

The Skrable et al. paper supports Courtney's 2008 conclusion and Berry's 2020 conclusion that the ocean is the source of the carbon that has caused most of the increase in atmospheric CO_2.

Meanwhile, IPCC's research went in a different direction. In 2011, an IPCC researcher published a paper that concluded outflow is proportional to level plus a constant term. This incorrect idea was derived not from physics. It was derived from a curve fit to data while using the incorrect assumption that IPCC's core theory is true.

Since this researcher was a climate model programmer, it is likely his incorrect formula is included in climate models.

Since 2011, many peer-reviewed IPCC papers use this incorrect formula for outflow. No IPCC paper has noticed this formula is incorrect.

The bottom line is IPCC's theoretical climate research operates in a different universe than standard physics. The reason appears to

41

be because the IPCC uses only "effects" scientists and no "cause" scientists.

Chapter 5 – The Scientific Method

"I would freely admit that on global warming we have crossed the boundary from news reporting to advocacy." - Charles Alexander, Time magazine science editor

"When the search for truth is confused with political advocacy, the pursuit of knowledge is reduced to the quest for power." – Michael Crichton, scientist, physician, author

5.1 Prosecution says Smith killed Jones.

Let's drop in on a fictitious federal murder trial in San Francisco.

Federal criminal court procedures are like the scientific method. The prosecution needs a unanimous vote by the jury. Only one "Not Guilty" vote defeats the prosecution's case.

Smith and Jones hit the same bar after work each day. They typically get into arguments. Smith even threatened Jones in front of the other patrons of the bar.

One day, neighbors found Jones's body in an alley. The autopsy showed Jones died from a 9mm gunshot wound at close range.

It did not take police long to identify Smith as a suspect. They got a search warrant to search Smith's home and found he owned a 9mm pistol that was recently fired.

The prosecution began its case against Smith by bringing in witnesses who testified Smith had once threatened Jones after a heated argument in the local bar. The prosecution showed Smith's 9mm pistol. Witnesses said this was likely the same type of pistol used to kill Jones, but they could not specifically say it was the same gun based on the evidence.

The case looked bad for Smith. The jury seemed to agree with the prosecution's argument that Smith was guilty.

Smith had smart defense attorney. The defense did not challenge any of the prosecution's evidence. Rather the defense showed evidence and witnesses who testified that Smith was in New York during and surrounding the day the prosecution showed that Jones was killed.

Case closed. Smith is innocent.

What happened?

The prosecution's case used circumstantial evidence because no one saw Smith shoot Jones.

The prosecution's theory was that Smith shot Jones. Every theory must be tested.

One prediction of the prosecution's theory is Smith had to be in the area at the time of the killing.

The defense proved this prediction of the prosecution's theory was wrong. Therefore, the theory was wrong.

The defense had no need to dispute any of the prosecution's evidence even if some of the evidence may have been disputable.

The defense did not need to find the real killer of Jones to prove Smith was innocent.

The scientific method works the same way. It does not require a replacement for a theory that you prove is wrong.

The IPCC claims "abundant published literature" shows, with "considerable certainty," that human CO_2 caused the rise in atmospheric CO_2 even though IPCC's core theory is wrong.

This IPCC claim is like the Smith-Jones jury concluding with "considerable certainty" that Smith shot and killed Jones even though Smith was 3000 miles away at the time.

5.2 The scientific method defined

John Kemeny earned his doctoral degree from Princeton in 1949. As a graduate student, he was Albert Einstein's assistant in mathematics. Einstein was working on unified field theory at the time. Kemeny later wrote:

> "People would ask - did you know enough physics to help Einstein?

> "My standard line was: Einstein did not need help in physics. But contrary to popular belief, Einstein did need help in mathematics. By which I do not mean that he wasn't good at mathematics. He was very good at it, but he was not an up-to-date, research-level mathematician. So, he needed an assistant for that, and, frankly, I was more up-to-date in mathematics than he was."

As a bonus, Kemeny learned the scientific method from Einstein. Years later, John Kemeny was chairman of both the Mathematics and Philosophy Departments at Dartmouth College.

The scientific method is a part of the Philosophy of Science. The scientific method is not an arbitrary set of rules. It is the only way to find truth in science.

As a Teaching Assistant at Dartmouth College, I worked on my M.A. in physics. I also took Kemeny's *Philosophy of Science* course. His course was so valuable to me that it should be required for all scientists.

I also took Kemeny's *Probability and Markov Chains*. That course became a key to my PhD thesis at the University of Nevada.

Figure 5.1 shows how Kemeny described the scientific method. The figure below is from his book *A Philosopher looks at Science*.

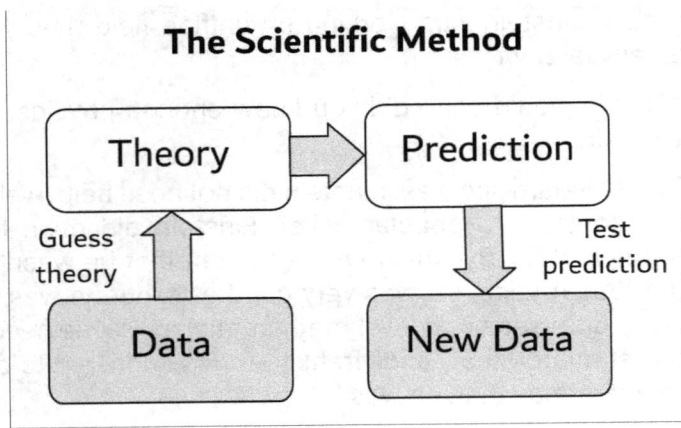

Figure 5.1. Outline of the scientific method.

A theory is a proposed explanation of how nature works. A theory must make predictions that we can test. We use the general definition of "theory" to include "hypothesis" and "idea."

A prediction connects a cause to an effect.

All theories begin and end with data. You use data to guess a theory that you think can predict new data. To test your theory, you make a prediction. Then you compare your prediction with new data. You do not test your prediction against the original data because that would be circular and not a valid test.

If your prediction is correct, your theory may be correct, but no number of successful predictions will prove your theory is correct. There is always the possibility that your next experiment will prove your theory is wrong.

The scientific method says no amount of evidence can prove a theory is true, but it takes only one incorrect prediction to prove a theory is false.

Einstein explained it this way (to paraphrase),

> Many experiments may prove me right, but it takes only ONE to prove me wrong.

46

Einstein understood that successful predictions do not prove a theory is right. But if a theory makes one false prediction, then the theory is wrong.

One test that proves a theory is wrong **outvotes** trillions of scientists who claim the theory is right.

Einstein said, "The case is never closed."

Richard Feynman was a Nobel Laureate in Physics. A YouTube video (see References) shows how he described the scientific method:

> "Now I'm going to discuss how we would look for a new law. In general, we look for a new law by the following process.
>
> "First, we guess it (audience laughter), no, don't laugh, that's the truth. Then we compute the consequences of the guess, to see if this law we guess is right, to see what it would imply.
>
> "Then we compare the prediction to nature, or to experiment or experience. We compare it directly with observations to see if it works.
>
> "If it disagrees with data, it's WRONG. In that simple statement is the key to science."

And we never, never, never, never adjust the data to fit our theory. But climate-effects scientists in the US government at NASA and NOAA have regularly modified past data to fit their false theories and models.

Science progresses by proving theories are wrong. If science does not test theories, it is not science. True science eliminates theories that make false predictions. The truth is in what is left after you eliminate bad theories.

5.3 Consensus is not science.

David brought a puppy to his Kindergarten class. After the kids played with the puppy, Nancy asked, "Is it a girl puppy or a boy puppy?"

After much discussion, the class could not decide. So, Johnny suggested, "Let's vote on it!"

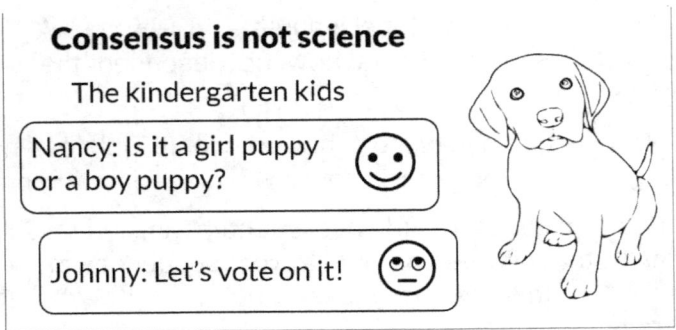

Figure 5.2. Votes do not determine scientific truth.

5.4 The scientific method and complex theories.

The scientific method applies to complex theories as well as to simple theories.

For example, a climate model is complex. It bundles many theories into one calculation. That bundle of theories is a new theory. We test this complex theory as we test simple theories. If a prediction is wrong, the theory is wrong.

In 1980, climate models began their predictions of future climate. Today, the models way over-predict temperature. Their predictions are wrong, so the models are wrong. The models do not even agree with each other.

5.5 The difference between science and religion

Kemeny defined the difference between science and religion as follows:

The scientific method applies only to our physical world. If we can measure it, it's science. If we cannot measure it, it's religion.

We cannot measure God and heaven. Therefore, they are part of religion. We can believe what we wish about religion because no one can prove our belief is wrong.

However, if a religion makes a claim about the physical world that we can measure, then we have entered the world of science. We can test our claim about the physical world and if it is wrong, we can prove it is wrong.

Chapter 6 – Cargo Cult Science

"We have to offer up scary scenarios about global warming … …each of us has to decide what the right balance is between being effective and being honest." - Stephen Schneider, Stanford University environmentalist

"The point of modern propaganda isn't only to misinform or push an agenda. It is to exhaust your critical thinking, to annihilate truth." – Gary Kasparov

6.1 The climate myth cripples America.

In South Africa in the 1960's, I saw children with their legs bound with ropes to make them grow up with crippled legs.

Today, in America, I see children with their minds bound with climate myths to make them grow up with crippled minds.

In a world where information abounds, people still believe in myths. Some myths do little damage. Other myths damage our economy, science, technology, national defense, and our minds. To make America great, there is no room for myths that do damage.

A survey, conducted from April 7 to 17, 2020, with an error margin of 3 percent, found 73 percent of Americans now believe our CO_2 causes climate change. 54 percent are "extremely" certain it is happening. Only 10 percent of Americans said human-caused global warming was not happening.

6.2 Cargo Cult Science

In 1974, Nobel physicist Richard Feynman addressed Caltech graduates. He warned the graduating class against using "Cargo Cult Science" in place of the scientific method.

Feynman described inhabitants of the South Seas who, during World War 2, watched American cargo planes land on their island on a real runway. They too wanted cargo planes to bring them good stuff.

Their cargo-cult theory was that runways attracted planes. So, they built a makeshift runway, set fires for runway lights, and donned bamboo headphones to re-create a setting like a nearby airport.

Even after World War 2 ended, they waited and waited, but planes never landed on their runway. They did not understand that their theory was wrong. Runways do not attract airplanes. Something else attracts airplanes.

Feynman explained how cargo-cult scientists, like the islanders, do almost everything right. They "follow all the apparent precepts and forms of scientific investigation, but they're missing something essential, because the planes don't land."

They did not understand the scientific method. Their prediction was wrong, so their theory was wrong.

The IPCC claims "multiple, independent lines of evidence show conclusively" their theory is true.

The cargo-cult people had multiple, independent lines of evidence to support their theory, but their theory was wrong.

The IPCC claims that observed temperature change, glacier shrinking, species change, wildfires, etc., prove their theory is true. But effects do not prove their cause.

Today, a whole generation of Americans follow cargo cult science because that is what they have been taught.

6.3 The Aztecs believed a climate myth.

The Aztecs needed rain for their crops. An Aztec high priest said, "If we offer sacrifices, God will send us rain."

Their theory was that human sacrifices will cause rain to occur. Of course, we know today that their theory was wrong.

Atop their pyramid-shaped temple, they lined up the people who would die that day.

They believed they had the power to change the climate.

If you were among those sacrificed, you would look up at the priest and watch his knife plunge into your chest. You may have even seen him raise your beating heart before you died. Then, they would cut off your head and roll it down the temple steps.

Many Aztecs believed so strongly in their theory, that they offered themselves to be sacrificed for the greater good of their people.

Eventually, after their priests rolled enough heads down their temple steps, rain came. It was inevitable, of course. But the Aztecs thought the rain proved their theory was correct. So, they continued their sacrifices.

They did not know what Aristotle said about confirmation bias. They did not know the Scientific Method.

The cost of their ignorance?

The cost was immediate death for those they sacrificed. The cost for the Aztec empire was eventual destruction because they wasted their human resources on a myth rather than using their human resources to solve problems.

6.4 Confirmation bias wrecks your thinking.

The scientific method is so subtle that few ever master it. The scientific method tells us to do the opposite of what our DNA, intuition, conscience, feelings, fears, and desires tell us to do.

Most people do exactly the opposite of the scientific method. They think a successful prediction proves their theory is right. But that is an illusion. They enhance their illusion when they reject data that proves their theory is wrong. That rejection creates confirmation bias.

Aristotle and Einstein taught us not to rely on data that reinforces our ideas. Such data do not prove our idea is true. Only data that negate our ideas lead us toward truth. Therefore, we must always seek data that contradicts our own ideas.

6.5 Effects do not prove their cause.

How often have you heard the phrase, "Climate change is real"?

To understand the phrase, you must understand what they mean by "climate change." In standard English, climate change means the climate changed. In alarmist English, climate change means "human-caused climate change is real."

So, the phrase, "Climate change is real" means to the alarmist, "we caused the climate change."

What prompted the statement? A weather event – say a wildfire – occurred. The hallucinated alarmist believes this effect was "unprecedented." Therefore, the alarmist concludes that this effect proves we caused the climate to change which, in turn, caused this effect to occur.

Aristotle described this logical error. It is called "affirming the consequent." It begins with the statement that A causes B. But they follow this with the logical error that, therefore, B causes A.

Figures 6.1 and 6.2 illustrate this logical error.

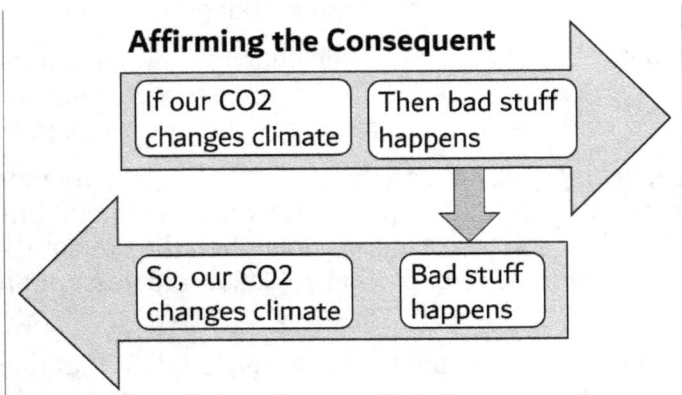

Affirming the Consequent

If our CO2 changes climate | Then bad stuff happens

So, our CO2 changes climate | Bad stuff happens

Figure 6.1. How alarmists use the logical error of Affirming the Consequent to conclude effects prove human CO_2 caused the effects.

"Bad stuff happens" is the consequent. But nature can cause bad stuff too.

However, the alarmist considers only one possible cause: humans. So, the alarmists affirm the consequent and reverse the logic to conclude we humans caused it.

Figure 6.2 shows the same logical error using something you are more familiar with.

Climate alarmism is built on this logical error.

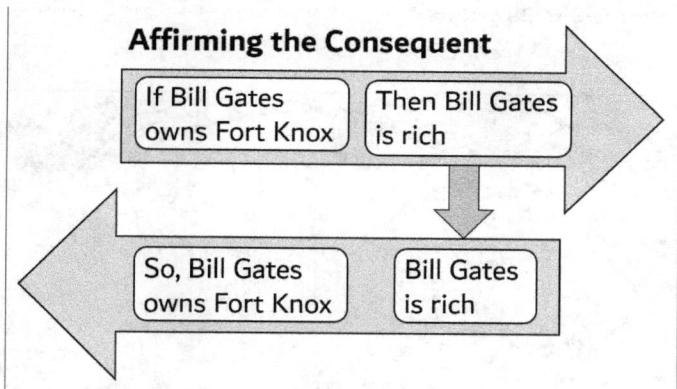

Affirming the Consequent

If Bill Gates owns Fort Knox → Then Bill Gates is rich

So, Bill Gates owns Fort Knox ← Bill Gates is rich

Figure 6.2. The same logical error becomes obvious when we include data we already know.

Politically, climate alarmists can ignore the cause.

Why? Because sloppy language used by the media has erased the distinction between natural-caused and human-caused climate change.

Today, the media only needs to claim climate change caused something bad to happen and most people will conclude human CO_2 caused it.

Effects are meaningless to the discussion of cause.

Suppose you have a headache. You popped a few aspirins and you still have a headache. So, you go to your medical doctor. Why?

You do not know what caused your headache. But your doctor is trained to find the cause of your headache. If your doctor did not do tests to determine the cause of your headache before giving you a prescription, it could be medical malpractice.

But climate alarmists do not test their theory of cause. So, they are guilty of climate malpractice.

Figure 6.3. Our ancestors thought they caused climate change. By permission of Rick McKee, Augusta Chronicle, February 1, 2007.

6.6 What is not a scientific argument.

The IPCC claims "extensive evidence" proves human emissions cause climate change. But extensive evidence is not a scientific argument and it is impossible to prove a theory is true.

Confirmation bias is not science. The IPCC rejects evidence that proves its core theory is wrong.

Appeals to authority have no bearing on scientific truth. Votes do not determine scientific truth and science is never settled. Aristotle and Einstein said so.

Ad hominem attacks on the messenger are not science.

Statements like "multiple, independent lines of evidence show conclusively," "vanishingly small," "thoroughly examined and tested" are cargo cult science.

6.7 California fires

On October 4, 2020, 60-Minutes host, Scott Pelley, interviewed some claimed climate experts. The program's purpose was to brainwash American voters to believe they must reduce their CO_2 emission to save the planet.

Pelley asked, "Are we too late to save the planet?" The question implies that we control the climate.

The program implied that effects prove their cause, which is invalid logic. But if you make the effects scary enough, you will convince most people.

Yes, it was hot and dry in California but that is local weather, not climate, and droughts are cyclical. California had bad droughts in 1840 and 1930.

A calculation I did for the State of California in 1990 predicted a severe drought from 2020 to 2030. Any good scientist can repeat this calculation using streamflow data and tree ring data.

The cycles in historical tree ring data predicted the 2020 drought. The 2020 drought has nothing to do with our CO_2.

There are hundreds of good scientists that 60-Minutes could have used to provide scientific balance to their program. But that would have conflicted with the goal of their program. The media are experts at confirmation bias. They select the people and data that support their goal.

6.8 CO_2 does not cause more severe weather.

Meteorologist Chuck Wiese wrote the following:

Much has been written about what the worldwide weather would be if human CO_2 emissions were changing the climate.

The 2012 paper by Francis and Vavrus introduced a big misconception. They claimed that as CO_2 warms the arctic, this would cause the jet stream waves to amplify and stall. This, turn, would increase the frequency of severe weather around the globe.

My 2016 paper demonstrates if atmospheric CO_2 changed the climate as Francis and Vavrus claim, then severe weather would decrease around the globe, not increase.

Today, atmospheric CO_2 continues to rise, and yet, there are no discernable changes in our weather patterns even though global temperatures are slightly warmer than they were 130 years ago.

This suggests the warmer temperatures are entirely natural and this natural warming trend is the dominant cause of the increase in CO_2.

Chapter 7 – The Climate Myth Origin

"A massive campaign must be launched to de-develop the United States. De-development means bringing our economic system (especially patterns of consumption) into line with the realities of ecology and the world resource situation." - Paul Ehrlich, Anne Ehrlich, and John Holdren, 1970.

"The great enemy of the truth is very often not the lie — deliberate, contrived and dishonest — but the myth — persistent, persuasive, and unrealistic." — President Kennedy

7.1 How climate alarmism began.

From the first Earth Day on April 22, 1970, to the Earth Summit in June 1992, climate alarmism was born and raised in politics.

Maurice Strong's lifetime goal was to transform the United Nations (UN) into a world government. From November 1970 until December 1972, Strong was Secretary-General of the United Nations Conference on the Human Environment.

In 1972, Strong founded and became the first Executive Director of the UN Environment Programme (UNEP).

Strong argued that rich Western countries had benefited by exploiting the earth's natural resources and, therefore, the Western countries must fund the poorer countries so their economies could catch up with America. President Obama supported this UN idea.

Under Strong's leadership, the 1972 United Nations Stockholm Conference made the environment an international agenda.

Strong commissioned the report by Barbara Ward and Rene Dubos, *Only One Earth: The Care and Maintenance of a Small Planet.* Their report promotes the Principles of the Stockholm Declaration which encourages people to safeguard natural resources and wildlife, share non-renewable resources, and indoctrinate the public to believe in UNEP's environmental cult.

7.2 The UN protects and promotes climate theory.

In 1978, Professor Bert Bolin of Sweden and his tiny band of meteorologists proposed that human CO_2 emissions cause the rise in atmospheric CO_2, and more CO_2 increases global temperature. Although he lacked scientific evidence, Bolin believed human CO_2 emissions could be harmful.

The International Council for Science (ICSU) and the UN World Meteorological Organization (WMO) sponsored the first World Climate Conference in Geneva in 1979. Bolin submitted a paper to the Conference. The WMO put Bolin's theory at the top of its agenda because a new disaster might help WMO get more funding.

Strong realized Bolin's idea that connected human CO_2 emissions to potential harmful results would support his goal of transforming the UN into a world government.

In October 1985, the UNEP and the UN World Meteorological Organization (WMO) sponsored the First International Conference on Climate Change in Villach, Austria. Bolin presented his theory with an urgent call to action.

The conference concluded that increasing concentrations of carbon dioxide could cause an historic rise in global temperature. This was a political conclusion, not a scientific conclusion. Bolin's idea was never tested with the scientific method.

To protect Bolin's theory from critique by the scientific community, Strong set up procedures that would block criticisms of Bolin's theory.

7.3 The Brundtland report, 1987

Strong was a member of the Brundtland Commission. The Brundtland report warned that human CO_2 could increase global temperature enough to harm agriculture, increase sea levels, flood coastal cities, and disrupt national economies.

The report called for a major global effort to curb human emissions of CO_2 and other greenhouse gases. It promoted the idea of "sustainability" as a possible solution to human-caused environmental problems.

7.4 The IPCC is born, 1988

In 1988, UNEP and WMO formed the UN Intergovernmental Panel on Climate Change (IPCC). More accurately, it is the "IP-on-CC".

The IPCC Charter states:

> "The role of the IPCC is to assess on a comprehensive, objective, open and transparent basis the scientific, technical and socio-economic information relevant to understanding the scientific basis of risk of human-induced climate change, its potential impacts and options for adaptation and mitigation."

There is nothing in IPCC's Charter about investigating the cause of climate change. The IPCC merely assumes our CO_2 causes climate change.

The U.S. government under President G.H.W. Bush was the main force in forming and funding the IPCC.

Under Strong's control, the IPCC appointed Bolin to be its first chairman and John Houghton, Bolin's supporter, to lead "Working Group 1" that would produce IPCC's climate reports.

Strong made IPCC's goal to produce reports that show human emissions cause climate change. IPCC's goal is NOT to find the real cause of climate change. The IPCC is a political organization, not a scientific organization.

Bolin's climate theory survived only because Strong made it a key part of UNEP and IPCC, and then protected this new IPCC climate theory from scientific criticism. This protection was critical because it gave Strong enough time to turn Bolin's climate theory into a political certainty.

61

Strong's protection of IPCC's climate theory still exists today. Many professional societies and professional journals will not publish scientific papers that disagree with IPCC's first theory. Strong masterminded the perfect crime.

7.5 America promotes the IPCC, 1988

In 1988, James Hansen, head of NASA's Goddard Institute for Space Studies (GISS), testified before a Senate committee chaired by Senator Tim Wirth. Senator Al Gore was on the committee. Hansen predicted the world was headed for a global warming disaster.

The media produced headlines across America and cover stories in Newsweek and Time. Senator Al Gore carried the climate change agenda in America.

In 1990, IPCC's First Assessment Report made global headlines, thanks to UN marketing power. It claimed human CO_2 caused global warming and warned that the world must reduce its CO_2 emissions by 60 percent immediately to save the planet.

Of course, there are no data that support those IPCC claims.

However, the environmentalists quickly adopted IPCC's climate claims because these climate claims supported and amplified their environmental agenda.

7.6 Earth Summit, 1992

In June 1992, Maurice Strong was Secretary General of the U.N. Conference on Environment and Development. He chaired the "Earth Summit" conference in Rio de Janeiro.

U.S. President G.H.W. Bush and 107 other world leaders attended the conference along with 20,000 climate activists and green lobby members. The UN and the US government paid all attendees' expenses.

Strong declared in his Summit speech,

"A shift is necessary toward lifestyles less geared to environmentally damaging consumption patterns. We may get to the point, where the only way of saving the world will be for industrialized civilization to collapse. Isn't it our responsibility to bring this about?"

Then Strong founded and chaired the Earth Council Alliance where he worked with Mikhail Gorbachev to create the Earth Charter which called for a

"... sustainable global society founded on the principles of respect for the Earth and life in all its diversity, economic and social justice, and a culture of peace and non-violence."

Strong declared,

"the real goal of the Earth Charter is that it will in fact become like the Ten Commandments."

Strong long supported global governance at the expense of national sovereignty. He said environmental mandates require the eventual dismantling of the power of the nation state:

"It is simply not feasible for sovereignty to be exercised unilaterally by individual nation-states, however powerful. It is a principle which will yield only slowly and reluctantly to the imperatives of global environmental cooperation."

"We need a system of global governance through which nations can cooperate and deal with issues they cannot deal with alone. The ultimate example is climate change."

In 1992, Al Gore claimed,

"Only an insignificant fraction of scientists deny the global warming crisis. The time for debate is over. The science is settled."

More accurately, the politics is settled but the science is not.

7.7 Strong becomes UN Under-Secretary General, 1997

In 1997, Strong became Under-Secretary General of the United Nations and served as a special advisor to UN head Kofi Annan. Strong used the UNFCCC to stage another mega-conference in Kyoto.

Strong was a leading architect of the 1997 Kyoto Protocol that set binding greenhouse gas reduction targets for 37 industrialized countries.

Strong inserted his long-term agenda into the Kyoto Protocol to commit 'developed' countries such as America to reduce CO_2 emissions and pay developing nations like China and India.

In 2000 and 2001, the Joyce Foundation, with Barack Obama on the board, granted $1.1 million to establish Gore's Chicago Climate Exchange (CCX) which made Strong a CCX director.

Strong died on November 28, 2015. The organizations he created to achieve his political goals are his legacy.

His goal was to turn the United Nations into a world government. He realized the idea that human CO_2 emissions increase global temperature, whether true or false, was a key to achieving his goal.

7.8 Environmentalism

Today's climate alarmism did not begin in the normal scientific process. It began in Strong's incubator that protected IPCC's climate theory from scientific critique. It flourished when environmental organizations adopted it into their programs.

After communism fell, environmentalism replaced communism. But it seems to have the same goal of world government as communism.

"Environmentalism" has a moral component. It alleges Man is destructive, unnatural, evil, and guilty of destroying the environment on this planet. Environmentalism is not a science. Its basic premise is nature is good and human is bad. IPCC reports assume the same moral view.

Environmentalism's moral assumption is embedded in Strong's remarks on behalf of the United Nations and in IPCC reports.

If you begin your climate study by assuming nature is good and human is bad, then you will conclude that natural CO_2 is good and human CO_2 is bad. If your environmental premise is that rising CO_2 is bad, then you will assume that human CO_2 caused it.

By contrast, physics is amoral. Physics tries to understand nature. Physics will get different answers to climate questions than environmentalism.

7.9 Political actions

Education from grade school through the highest levels must teach rigorous logic and critical thinking. Students must learn to doubt fashionable theories, and to distrust all "hop-on-the-bandwagon" dogma.

US President Trump tweeted on July 10, 2020:

> "Too many Universities and School Systems are about Radical Left Indoctrination, not Education. Therefore, I am telling the Treasury Department to re-examine their Tax-Exempt Status and/or Funding, which will be taken away if this Propaganda or Act Against Public Policy continues. Our children must be Educated, not Indoctrinated!"

To this day, The UN, all governments (Yes, even the US government), all scientific organizations, all schools, colleges, and universities, and all the major media continue to protect the IPCC climate theory from scientific criticism. If you are in a space they control, they will not allow you to ask any questions that suggests their cult belief is wrong.

Their globalist call "to address climate change" is a diversion from much more important problems. For example, a much greater concern for many countries is protection against an EMP attack.

In 2008, America's EMP Commission warned that America must ensure the safety of its power and information grid against an EMP attack. An electromagnetic pulse (EMP) is a short burst of electromagnetic energy, natural or manmade.

The Obama–Biden administration did nothing to protect Americans for eight critical years.

In 2016, the U.S. Government Accountability Office ("GAO") reported that the federal government had not implemented its recommendations to prevent massive damage by an EMP attack.

The EMP Commission warned Congress that an EMP attack would "shut down America's electric power grid." Within a year, 90 percent of Americans would be dead.

In March 2019, US President Trump signed an executive order that instructed federal agencies to strengthen America's infrastructure against EMP attacks. It was the first order of its kind to establish a comprehensive policy to improve resilience to EMPs.

Climate change can take decades and it will not kill us.

An EMP attack can take seconds to knock a country into the stone age and kill 90 percent of the people within a year.

Chapter 8 – The Trillion Dollar Fraud

IPCC's climate theory "is the greatest and most successful pseudoscientific fraud I have seen in my long life as a physicist." - Harold Lewis, Emeritus Professor of Physics, University of California, in his resignation letter to the American Physical Society, emailed on October 8, 2010.

"Future generations will wonder in amazement how the 21st century world went into hysterical panic over a globally averaged temperature increase of a few tenths of a degree - on the basis of gross exaggerations of highly uncertain computer projections and implausible chains of inference - contemplated a roll-back of the industrial age." - Richard Lindzen, MIT Professor of Meteorology

America paid the IPCC and its supporting scientists a trillion dollars, and they did not get basic physics correct. The IPCC has no use for "cause" scientists because it assumes its core theory is true.

The "effects" scientists tout their "peer-reviewed" papers that assume IPCC's core theory is true before they claim to prove IPCC's core theory is true, in the most blatant example of mass circular reasoning in history.

Science is not UN's goal. The UN uses pseudoscience to achieve its political goals. Their goal is to control you and your country.

This book shows you how to prove the IPCC climate claims are a fraud.

You do not have to be a scientist to use this proof. This proof is not "an opinion" as the alarmists may claim. This proof can win in court.

Here are the steps to prove the IPCC is wrong that you learned in this book:

8.1 IPCC's core theory is the natural CO_2 level stayed constant at 280 ppm before and after 1750.

1. Data show the CO_2 level rose to 410 ppm by 2020, an increase of 130 ppm.

2. The IPCC assumes its core theory is true, which forces the conclusion that human CO_2 caused all the increase above 280 ppm.

3. IPCC agrees that human CO_2 emissions are less than 5 percent of natural CO_2 emissions.

4. How can less that 5 percent of all CO_2 emissions cause 32 percent of the CO_2 in the atmosphere? Answer: It can't.

8.2 Multiple lines of evidence prove IPCC's core theory is wrong.

1. Ice core data prove natural CO_2 caused the CO_2 increase.

2. Direct CO_2 data prove CO_2 was much higher than 280 ppm before 1750.

3. Leaf stomata data prove CO_2 was much higher than 280 ppm before 1750.

4. Statistics prove human CO_2 is not the primary cause of the increase in CO_2.

5. IPCC's human carbon cycle is not consistent with its own natural carbon cycle. This is a basic physics error.

6. Inspection shows IPCC's human carbon cycle is based on IPCC's invalid assumption that its core theory is true.

8.3 A simple physics carbon cycle model replicates IPCC's data for its natural carbon cycle.

1. This model easily calculates the true human carbon cycle that is compatible with IPCC's natural carbon cycle.

2. The true human carbon cycle shows human CO_2 has added only 33 ppm to the CO_2 level as of January 2020.

3. The IPCC 20-percent error bounds of this 33 ppm are 24 ppm and 48 ppm, with these bounds being improbable.

8.4 The true human carbon cycle shows:

1. If human CO_2 emissions were to stop in 2020, the human-caused 33 ppm increase would decrease to 16 ppm in 20 years and to 10 ppm by 2100.

2. Natural CO_2 caused about 100 ppm of the CO_2 increase since 1750.

3. Human CO_2 does not stick in the atmosphere for thousands of years as IPCC claims.

4. Human CO_2 is not a threat to the planet.

5. Stopping all human CO_2 emissions cannot reverse nature's 100 ppm or stop nature from increasing the level of CO_2.

8.5 This proof that IPCC's core theory is false means:

1. All peer-reviewed scientific papers that assume, openly or secretly, that IPCC's core theory is true are invalid.

2. There is no basis for climate laws, climate regulations, climate treaties, climate brainwashing, and climate environmentalism.

3. There is no basis to continue funding the IPCC or any research based that assumes IPCC's core theory is true.

4. There is a basis for removing all textbooks and literature that claim IPCC's core theory is true.

5. There is a basis for finally telling people the truth about climate change.

8.6 What would Dorothy do?

The cyclone takes you, Dorothy, and Toto to the Land of Oz.

There you meet the Wicked Witch of the UN who promotes the climate fraud. Toto pulls back the curtain and reveals the climate fraud.

This book is Toto.

Our job is to use climate truth to kill the Wicked Witch of the UN.

We may have to give brains to scarecrows, courage to lions, and hearts and a few drops of oil to tinmen.

When we finish, we can click our heels three times and return home ... to Kansas or wherever we live.

And when Dorothy returns home, she will go to this link:

https://edberry.com/climate-miracle/

for more information, references, discussions, and videos about this book.

On that link, she will find this clickable link:

https://www.amazon.com/dp/B08L5Z3LMR/ref=sr_1_1

where she will write an excellent "customer review" of this book, and she recommends you do the same.

The *Climate Miracle* is nature controls the climate.

Let's Roll!

> **Alarmists die a thousand deaths.**
>
> **The wise die only once.**

References

All *Climate Miracle* References are posted here:

https://edberry.com/climate-miracle/

Acknowledgements

The author thanks Howard "Cork" Hayden, Fryar Calhoun, Peter Morgan, and Valerie Berry for your comprehensive edits and suggestions.

Thanks to all who wrote an *Enthusiastic Recommendation*.

Thanks to all who provided edits and suggestions: Joseph Apple, Robert Bugiada, Richard Courtney, Joan Donohoe, Joel Glassett, Alan Greer, Fred Hammel, David Houghton, Cees De Jong, Kenneth Kok, Susan Lake, Daniel Lamanuazzi, Pamela Matlack-Klein, Max Polo, Gordon Raboud, Dennis G Sandberg, John Shanahan, Michael Spencer, Larry Vardiman, Duncan White, and David Whitmore.

Thank you Jerry Molen for your Foreword and Ed Rush for your business coaching.

Thanks to all who reviewed the author's Preprint #3, which is the scientific basis of *Climate Miracle*: Hermann Harde, Richard Courtney, Nils-Axel Morner, Chuck Wiese, Gordon Fulks, Gordon Danielsen, Larry Lazarides, John Knipe, Ron Pritchett, Alan Falk, Leif Asbrink, Mark Harvey, Case Smit, Stephen Anderson, and Chic Bowdrie.

Enthusiastic Recommendations

Climate Miracle is an excellent, up-to-date overview of the current "climate change" controversy. It systematically exposes the fallacies behind the climate alarmists' talking points, while revealing numerous fabricated assumptions driving the IPCC agenda to demonize CO_2. It also clearly explains the false logic used to exaggerate the human versus natural CO_2 concentrations. If you seek climate logic versus climate lunacy, *Climate Miracle* is an exceptional resource.

John D. Shewchuk, CCM
Eosonde Research Services, LLC.
Lt Col, USAF, Retired (Advanced Weather Officer)
Signatory to the CLINTEL's World Climate Declaration

Dr. Berry is our climate expert. *Climate Miracle* is essential reading.

Paul E Vallely, MG US Army (Ret)
Chairman - Stand Up America US
Chairman - Legacy National Security Advisory Group
Founding Member - Citizens Commission on National Security
Web Site: www.standupamericaus.org

I have known Ed (Dr. Berry) for more than a decade and have found him to be one of the more informed individuals on climate cause and effect than anyone I have come in contact with. His straightforward, non-nonsense approach has always been a breath of fresh air in a world where political correctness permeates the very fabric of science instead of the logic and reasoning is deserves. Another well laid out publication that should be well received in the community of his peers. Bravo!!

Glenn Wehe
Montana

Ed Berry has performed an extraordinary service to the climate community, science in general, America, and the world by identifying the underlying errors the IPCC has made in its fraudulent claim that human-caused climate change is an existential threat to life on this planet.

Climate Miracle is a popular treatise on the deceptive use of science, so called, to bring about political change in a world fearful of an increase in atmospheric carbon dioxide. The title of his book highlights one critical error the climate alarmists themselves don't realize - the presence of a supernatural agent required to separate natural carbon dioxide molecules from human-created molecules in the atmosphere needed in their theory.

Without this fictitious agent that Berry calls a demon, the IPCC calculations are just so much gibberish. James Clerk Maxwell in 1867 suggested a similar such demon to refute the idea in thermodynamics that a perpetual motion machine was possible. With similar logic Berry refutes the IPCC prediction that catastrophic climate change is imminent if such a demon isn't present.

Berry has done his homework using theoretical and data-based analyses to demonstrate the fallacies and the shear lack of proper scientific protocol used in their reports on global climate change to be published in peer-reviewed journals elsewhere. This book is a well-written attempt to summarize the conclusions and consequences of misusing the scientific method for frightening the public into submission to an authoritarian political agenda.

Larry Vardiman, PhD
Colorado State University
Cloud Physicist
Paleoclimatologist

Thank you, Dr. Berry, for writing this book on climate. It takes one of the monumental confusions of our day and puts in terms we can all understand. It gives us hope at a space in time when climate alarmists bring us none.

Susan Lake
Montana

This is a wonderful book, technically correct, lucidly presented, and easily understood by all. It fills a great need to counter environmental brainwashing the world is receiving from the mass media today. I will buy it from Amazon and give it to as many people as I can.

Cecil Joe Tomlinson

Dr. Ed Berry is a great American climate physicist, educator, and citizen. Everything he does, he does well.

His October 2020 book, *Climate Miracle* starts out with a great title and photo of the sky, land, and water in Montana. This book addresses climate science, scientific errors and sinister intentions of the United Nations and their Intergovernmental Panel on Climate Change, the Scientific Method versus Religion, how the man-made global warming myth was born, and straight-forward conclusions.

Most important is to understand that man-made climate change alarmism intends to deny people access to fossil fuels and all their benefits. It is also to establish a one-world dictatorship government that will impoverish billions of people. Everyone should read this book and educate their elected officials.

John Shanahan

Al Gore's inconvenient lies in 2006 were enough to convert me sufficiently to skepticism about the then emerging propaganda of man-made dangerous global warming that I co-founded the New Zealand Climate Science Coalition and have remained its honorary secretary and webmaster ever since.

In that time, I think I have read all of the main literature opposing the alarmism now known as "climate change" because of the cessation of any appreciable warming after the expiry of the late 20th century extreme El Nino.

I thought I knew all the arguments necessary to combat alarmist propaganda, until I saw a draft of Dr Ed Berry's short treatise, *Climate Miracle*.

Eminent physicist, Dr Berry has introduced me to two new and compelling arguments demolishing the claims of the UN IPCC and its tax-grabbing minions and exposing them as a litany of lies.

Demolition No 1:

Dr Berry identifies two opposing camps of so-called climate scientists:

(i) the "effects" camp, comprising taxpayer-funded lackeys who have forged what they claim is a "consensus" that emissions by human and animal kind of the trace gas Carbon Dioxide (CO_2) will cause Earth's temperatures to soar and the seas to rise to swamp low-lying land.

They base their consensus on computer projections into which they have inserted speculative numbers designed to yield the high readings they bleat about, for which their political and would-be world governance wannabes reward them with huge financial grants, mostly from our tax payments.

(ii) the "causes" camp, comprising thousands of independent and retired scientists still searching for real-world answers to what makes our Earth's weather tick in the natural cycles we have known and adapted to successfully for millennia.

Dr Berry's explanation of the motives and modus operandi of these two camps easily compels a finding of support for the integrity of the "causes" camp.

Demolition No 2:

Dr Berry identifies three core theories on which the IPCC relies for support for its alarmist propaganda, saying that proof that just one of those three theories is false is enough to render IPCC's complete theory false.

Dr Berry then demonstrates that IPCC's core theory is false and blows the IPCC case to smithereens. Buy the book to learn more about the three bogus theories. It's the first time I've seen this argument advanced with such compelling cogency.

Terry Dunleavy MBE
Hon Secretary, New Zealand Climate Science Coalition (www.climatescience.org.nz)
New Zealand Ambassador for the Climate Intelligence Foundation (www.clintel.org)

Your work is extremely important and enlightening. My Chemical engineering training helps me understand your material balances and I agree that your science is simple, profound and must be used to guide future policies responding to climate change.

I have studied climate data as a curious engineer and found that I disagree with two IPCC narratives.

IPCC's narrative is "observed seasonal variations of atmospheric carbon dioxide concentrations are caused by vegetation seasonal growth and decay". I believe that seasonal variations of carbon dioxide are better explained by "global/seasonal changes in absorption of atmospheric carbon dioxide by sea water."

IPCC's narrative is that "non-GHG factors from human activity on earth's surface is so small that these effects can be ignored in climate models". I and Chinese climate scientists see data and logic that indicates "non-GHG factors may be a very large part of any recent warming trends".

Dr Berry's conclusion that human carbon dioxide emissions is not an important driver of climate change is a game changer.

All these findings that invalidate IPCC's narratives have something in common. They are ALL IGNORED (Not disputed!) by mainstream climate scientists. You would think that IF they were WRONG, climate scientists would simply point out the errors in science, data, and logic.

I believe that the US should create a climate change commission made up of thinking engineers and scientists that have NO self interest in the climate change business/politics to evaluate important theories/work such as Dr. Berry's that challenges IPCC narratives to make sure that future policies are based on sound science.

Ed Sebesta

Your effort in publishing data concerning matters that many of us believe but don't have the credibility to publish is most welcome! Good luck! Will promote your book widely.

Rosemary Falcon

Chapter 7 is most enlightening on the creation of the political movement, and it exposes the underlying intent of the sham.

Bob Denson

With only a BS degree, and being more in tune with public/media/education relations than hard-core academic facts, the fundamental chemistry of CO_2 convinces me that our earth's atmosphere does not and cannot distinguish between natural or human-expelled CO_2; it's all the same chemical wherever you find it.

Consequently, it is sobering, scientifically and spiritually, to realize that our intelligently-designed atmosphere has a built-in method for balancing CO_2 to the levels both needed and tolerable for earth's living organisms, just as it has been since "the beginning."

Bob Webster

Thanks Dr Ed, your logic and simplicity in explaining such complex matters are outstanding. Thanks for debunking lies and pseudoscience of bureaucrats and fake scientists.

Time for truth & honesty to take the lead in climate science!

Max Polo
Technical Manager, Energy Division, Cimolai SpA – Italy

What an absolutely compelling demolition of the IPCC's nonscience (spelling intentional)! It's by far the best I've ever read, and over the 30-odd years that I've been a skeptic (as everyone trained in science and engineering should always be), I've read quite a few attempts, including some that I've written myself.

Peter J. Morgan B.E. (Mech.), Dip. Teaching, Honorary Chairman and Chief Executive Officer, Environomics (NZ) Trust, Consulting Forensic Engineer, Marine Designer, Technical Writer, Sub-editor & Technical Editor

---

Thank you, Ed, for this educational book, that should not be missing in any school or university library.

Cees De Jong

Congratulations and best wishes with your brilliant book.

David Whitmore
Waterlooville UK

Amazing, superb book. A shot at the heart of the beast. In only a few pages you will be absorbed and well informed to fight the good fight with the weapons of climate realism.

Rodrigo Penna-Firme, PhD

It's so interesting that the same theme is used both in the promotion of the 'climate change' fraud and also presently in the COVID-19 situation, and that same theme can be summed up in one word: 'FEAR'!

Fear is the ingredient that has influenced politics going way back in human history where someone who is intent on mastery influences the generally-unthinking masses to follow their lead. (And there are several classic examples of this in the 20th Century – Adolf and Josef come to mind! – and look where that got us all.

Ed, well done with your basic explanations of physics, and the idea of a 'debate' is a good one.

Michael Spencer

This is clear, logical, and easy to understand. I would give this book to anyone who is worried about the climate scam. My only reservation is that most of the people I know who believe in human caused climate change are so indoctrinated that they would not be willing to read it. I particularly appreciated the warning in the last chapter about the slow encroachment of communism in the takeover of our institutions.

James Pearce

A very well-reasoned document. You get over the point that there needs to be solid evidence to back up a guess. The climate models are a joke. I would like to make the point that CO_2 is only 1 molecule in 2500. Consider this brave (May 12th, 2012) admission by German Physicist and meteorologist Klaus-Eckart Plus:

"Ten years ago, I simply parroted what the IPCC told us. One day I started checking the facts and data—first I started with a sense of doubt but then I became outraged, when I discovered that much of what the IPCC and the media were telling us was sheer nonsense, and was not even supported by any scientific facts and measurements. To this day I still feel shame that as a scientist, I made presentations of their science without first checking it. ...scientifically it is sheer absurdity to think we can get a nice climate by turning a CO_2 adjustment knob".

Congratulations, and best of luck on Amazon.

Ian Storey

Quick and concise read. Enjoyed it. Good info that I sort of knew but couldn't explain it like you did.

Jimmy Arnold

A lone, self-funded scientist in a small Montana town does research that bright high schoolers can duplicate, proving the carbon cycle is a massive fact of nature and we humans are a tiny part of that natural pattern.

Fran Tabor

Your book is excellent. Since it is directed largely at the scientifically-challenged among us, I recommend adding an argument that I have used which is not scientific at all but just common sense, an approach is very effective for those without scientific background:

If the people urging laws to reduce CO_2 really think the situation is so desperate, why is their only solution one that even they themselves agree will only reduce world temperatures by .02 degrees in 50 years?

(A fact most people are unaware of since it is carefully ignored in mainstream media. You could show the math and list the authorities who have admitted to the .02 degrees figure.)

Why don't they urge other solutions easily implemented right now, such as not rebuilding on flooded lands and restricting further development along ocean shores, solutions that would be effective whether global warming is happening or not?

And why are they silent about the imminent issue (within 20 years) of future recycling for the millions of solar panels and wind turbines they want to blanket across our land, equipment made of highly poisonous "rare earth" materials?

Bonnie Chandler

Thanks for this excellent resource. I loved your logical approach to cause and effect. It's just not possible to reverse these!

Brinsley Jenkins

Excellent explanation of the climate hoax between Globalist Gore and Wisdom Will followed by concise physics data!

John A. Bird

Thanks Ed, here's hoping your book gets the circulation our country desperately needs.

Dennis G Sandberg

Thanks for the great work you are doing.

Philip Mulholland

You certainly have a large number of very intelligent people who are interested in the success of your book. Clearly respecting their inputs, no doubt your published work will be much improved, and I look forward to purchasing a copy.

Richard McFarland

Ed, I've followed your development of the physics you used to show the flawed assumptions made about atmospheric CO_2 from the IPCC that attribute all of it to human emissions. The development of your arguments is nothing less than phenomenal to their conclusion that now include all of the carbon reservoirs on the earth.

This book is a beautiful summary of these developments and should be read by all to avoid the disastrous public policy mistakes that climate hysterics are making and want to force upon the public due to the flawed science that is being used by the IPCC.

Congratulations on your efforts and hard work, Ed. I hope this book heads for a best seller category to properly educate the citizens of this country about climate and the true non problem human CO_2 emissions really are to it and richly rewards you for your efforts to publish sound science.

Chuck Wiese
Meteorologist
Professional Pilot

About the Author

Dr. Ed Berry is CEO of Ed Berry, LLC, in Bigfork, Montana.

He received his BS in Engineering from the Caltech, his MA in Physics from Dartmouth, and his PhD in Physics from the University of Nevada. Berry's PhD thesis is still cited as a breakthrough in cloud physics and numerical modeling. His model is used in biology, geology, and cosmology.

Berry is an American Meteorological Society Certified Consulting Meteorologist, and a pilot, with glider, power, and instrument ratings. Once, he flew a sailplane into a Nevada storm and then made a zero-zero landing on a Nevada dry lake in a severe dust storm.

He was chief scientist for Nevada's Desert Research Institute airborne research facility. He led pioneering research flights into Alberta hailstorms and Sierra Nevada mountain storms. He designed the first airborne, earth-referenced radar that predated modern GPS instruments.

He was the only civilian in a Department of Defense top-secret weather modification project in southeast Asia.

Berry was the National Science Foundation's Program Manager for Weather Modification, where he managed the Metropolitan Meteorological Experiment (METROMEX), National Hail Research Experiment, and other university research projects.

In his private business, he showed the Federal Aviation Administration how to prevent airline crashes caused by severe downdrafts.

He performed the Southern California Desert wind-energy study for the California Energy Commission. He identified Altamont Pass and Tehachapi Pass as excellent wind resources.

Berry used his courtroom numerical model to successfully defend the accused of murder against California Attorney General. He modeled human body responses to heat and humidity inside a box inside a van on a trip from northern to southern California on a hot summer day. His model won the 1993 Microsoft Windows World Open "People's Choice Award."

The University of Nevada Alumni Association gave Berry its Professional Achievement Award.

Berry led Montana's successful fight against the *Our Children's Trust* climate lawsuit.

As a graduate student at Nevada, his athletic performance earned him a membership in the elite Sigma Delta Psi national athletic honorary.

He and his wife Valerie as crew won sailing Gold Medals in the 1974 Canadian Olympic Regatta in Kingston, Ontario. They also won US national and North American championships in their high-performance 16-foot centerboard sailboat.

He has placed in the USA Top-10 age-group run-bike-run biathlons and track events. He holds age-group world records in Concept2 rowing.

Dr. Berry speaks, teaches, advises, and consults on climate change.

Climate Miracle

There is no climate crisis. Nature controls climate.

Edwin X Berry, PhD, Physics
Certified Consulting Meteorologist

Ed Berry, LLC
439 Grand Dr #147
Bigfork, Montana 59911
ed@edberry.com

Dr. Berry speaks, teaches, advises, and consults on climate change. Go to:

https://edberry.com/climate-miracle/

to:

- See the References for this book
- Join the conversation about this book
- Join the *Climate Miracle* Membership

Hiring Dr. Ed to speak

Hey there! It's Ed again. If you host events (or know someone who does) let's talk.

One of my passions is speaking and I would love to bring an inspiring message to your corporation, association, or group.

I can also present a several hour course on this book.

Of note, I do this as a business, so I do not travel to speak because it's the "right thing to do." But if you have a real budget and you want to hire a speaker who is a real climate scientist, let's talk.

https://edberry.com/climate-miracle/

Hiring Dr. Ed for Consulting

One of the fastest ways to decide how to "address" climate change is to get advice from an expert in the cause of climate change.

My skill is to show business and political leaders how to understand the debate about climate change without having to learn climate science.

Most business and political leaders do 100 things and feel guilty about the other 100 things they don't have time to do.

When we're done talking, you'll have the climate information you need to build the climate plan.

I am expensive and you probably can't afford me. But for the right business or political issues, what I know may save you a lot of time and money.

https://edberry.com/climate-miracle/

Social Media

LinkedIn: edwin-berry

Instagram: dr.edberry

WeMe.com: EDWINBERRY

Tell your social media friends about this book.

CPSIA information can be obtained
at www.ICGtesting.com
Printed in the USA
LVHW081541231120
672477LV00050B/3763